AF831020

COMPUTATION OF GENERALIZED MATRIX INVERSES AND APPLICATIONS

COMPUTATION OF GENERALIZED MATRIX INVERSES AND APPLICATIONS

Ivan Stanimirović, PhD

APPLE ACADEMIC PRESS

Apple Academic Press Inc.
3333 Mistwell Crescent
Oakville, ON L6L 0A2
Canada

Apple Academic Press Inc.
9 Spinnaker Way
Waretown, NJ 08758
USA

© 2018 by Apple Academic Press, Inc.

Exclusive worldwide distribution by CRC Press, a member of Taylor & Francis Group

No claim to original U.S. Government works

International Standard Book Number-13: 978-1-77188-622-2 (print)

International Standard Book Number-13: 978-1-315-11525-2 (ebook)

All rights reserved. No part of this work may be reprinted or reproduced or utilized in any form or by any electric, mechanical or other means, now known or hereafter invented, including photocopying and recording, or in any information storage or retrieval system, without permission in writing from the publisher or its distributor, except in the case of brief excerpts or quotations for use in reviews or critical articles.

This book contains information obtained from authentic and highly regarded sources. Reprinted material is quoted with permission and sources are indicated. Copyright for individual articles remains with the authors as indicated. A wide variety of references are listed. Reasonable efforts have been made to publish reliable data and information, but the authors, editors, and the publisher cannot assume responsibility for the validity of all materials or the consequences of their use. The authors, editors, and the publisher have attempted to trace the copyright holders of all material reproduced in this publication and apologize to copyright holders if permission to publish in this form has not been obtained. If any copyright material has not been acknowledged, please write and let us know so we may rectify in any future reprint.

Library of Congress Control Number: 2017953052

Trademark Notice: Registered trademark of products or corporate names are used only for explanation and identification without intent to infringe.

Library and Archives Canada Cataloguing in Publication

Stanimirović, Ivan, author
Computation of generalized matrix inverses and applications / Ivan Stanimirović, PhD

Includes bibliographical references and index.
Issued in print and electronic formats.
ISBN 978-1-77188-622-2 (hardcover).--ISBN 978-1-315-11525-2 (PDF)
1. Matrix inversion. 2. Algebras, Linear. I. Title.
QA188.S73 2017 512.9'434 C2017-905903-3 C2017-905904-1

CIP data on file with US Library of Congress

Apple Academic Press also publishes its books in a variety of electronic formats. Some content that appears in print may not be available in electronic format. For information about Apple Academic Press products, visit our website at **www.appleacademicpress.com** and the CRC Press website at **www.crcpress.com**

CONTENTS

Dedication

To Olivera

ABOUT THE AUTHOR

Ivan Stanimirović, PhD
Assistant Professor, Department of Computer Science, Faculty of Sciences and Mathematics, University of Niš, Serbia

Ivan Stanimirović, PhD, is currently with the Department of Computer Science, Faculty of Sciences and Mathematics at the University of Niš, Serbia, where he is an Assistant Professor. He formerly was with the Faculty of Management at Megatrend University, Belgrade, as a Lecturer. His work spans from multi-objective optimization methods to applications of generalized matrix inverses in areas such as image processing and restoration and computer graphics. His current research interests include computing generalized matrix inverses and its applications, applied multi-objective optimization and decision making, as well as deep learning neural networks. Dr. Stanimirović was the Chairman of a workshop held at 13th Serbian Mathematical Congress, Vrnjačka Banja, Serbia, in 2014.

PREFACE

Computation of Generalized Matrix Inverses and Applications offers a gradual exposition to matrix theory as a subject of linear algebra. It presents both the theoretical results in generalized matrix inverses and the applications. The book is as self-contained as possible, assuming no prior knowledge of matrix theory and linear algebra.

The book first addresses the basic definitions and concepts of an arbitrary generalized matrix inverse with special reference to the calculation of $\{i,j,\ldots,k\}$ inverse and the Moore–Penrose inverse. Then, the results of LDL^* decomposition of the full rank polynomial matrix are introduced, along with numerical examples. Methods for calculating the Moore–Penrose's inverse of rational matrix are presented, which are based on LDL^* and QDR decompositions of the matrix. A method for calculating the $A_{T,S}^{(2)}$ inverse using LDL^* decomposition using methods is derived as well as the symbolic calculation of $A_{T,S}^{(2)}$ inverses using QDR factorization.

The text then offers several ways on how the introduced theoretical concepts can be applied in restoring blurred images and linear regression methods, along with the well-known application in linear systems. The book also explains how the computation of generalized inverses of matrices with constant values is performed, and covers several methods, such as methods based on full-rank factorization, Leverrier–Faddeev method, method of Zhukovski, and variations of partitioning method.

Key features of the book include:

- Provides in-depth coverage of important topics in matrix theory and generalized inverses of matrices, highlighting the Moore–Penrose inverse.
- Requires no prior knowledge of linear algebra.

- Offers an extensive collection of numerical examples on numerical computations.
- Describes several computational techniques for evaluating matrix full-rank decompositions, which are then used to compute generalized inverses.
- Highlights popular algorithms for computing generalized inverses of both polynomial and constant matrices.
- Presents several comparisons of performances of methods for computing generalized inverses.
- Includes material relevant in theory of image processing, such as restoration of blurred images.
- Shows how multiple matrix decomposition techniques can be exploited to derive the same result.

CHAPTER 1

INTRODUCTION

In mathematics and computer science, an algorithm is a step-by-step procedure to make calculations. Algorithms are used for calculation, data processing, and automated reasoning. More specifically, the algorithm is an effective method expressed as a final list of defined instructions. Starting with the initial steps and inputs (which may be empty), instructions describing the calculations, which, after execution, lead to a finite number of successive steps, giving output in the last step. Complexity of an algorithm can be described as the number of primitive operations or basic steps that are necessary to make for some input data (more analysis on algorithms can be found in Ref. [10]).

Matrix computations are used in most scientific fields. In every branch of physics—including classical mechanics, optics, and electromagnetism—matrix computations are used for the study of physical phenomena. In computer graphics, matrices are used for designing three-dimensional images on a two-dimensional screen. In theory, probability and statistics, stochastic matrix is used to describe sets of probability. Matrices have a long history of application in solving linear equations. A very important branch of numerical analysis is dedicated to the development of efficient algorithms for matrix computations. The problem is well-known for centuries, and today it is a growing field of research.

In linear algebra, matrix decomposition represents the factorization of a matrix into a canonical form. Method of matrix decomposition simplifies further calculations, either theoretical or practical. There are algorithms that are made for certain types of matrices, for example, rarely possessed or near-diagonal matrix. There are many methods for the decomposition of a matrix, where each is used in separate classes of problems. Decomposition related to solving a system of linear equations includes *LU decomposition, LU reduction, Block LU decomposition, rank factorization, Cholesky decomposition, QR decomposition,* and *singular value decomposition.*

Symbolic calculation is a concept that makes use of computer tools for transformation of mathematical symbolic expressions. It is most often used for the calculation of explicit results without numerical errors. Therefore, symbolic calculation is always applied to solve conditioned problems. Methods of symbolic computation have been used in manipulating complicated expressions of several variables, which may be rational functions or polynomials of one or more variables. There are several different software packages for computer algebra (Computer Algebra Software) that support symbolic calculation, such as `MATHEMATICA`, `MAPLE`, and `MATLAB`.

Traditional programming languages are procedural languages (`C`, `C++`, `Fortran`, `Pascal`, `Basic`). A procedural program is written as a list of instructions that are executed step-by-step [90]. Computer programs in procedural languages such as `C`, can be used in the calculation, but they are limited to understanding more complex algorithms, because they give little information about the intermediate steps. The user interface for such programs typically requires definitions of user functions and data files corresponding to a specific problem before the running time [4]. Many researchers use the possibility of developing "rapid-prototype" code to test the behavior of the algorithm before investing effort in the development of the algorithm code in a procedural language. Access to MATHEMATICA has great advantages in research over procedural programming available in MAPLE-u and in procedural languages [90]. MATHEMATICA leaves several programming paradigms: object-oriented, procedural, symbolic, and functional programming [90]. The goal is to develop algorithms suitable for implementation in both MATHEMATICA and procedural programming languages.

Decompositions based on eigenvalues and related concepts are: *Eigen decomposition, Jordan decomposition, Schur decomposition, QZ decomposition,* and *Takagi's factorization.*

Before we move on to the explanations of the basic structural characteristics of the matrix, let us first see one of the more general divisions of the structural matrix. This is a very important class of matrices, which are ubiquitous in algebra and numerical calculations in various scientific disciplines such as electrical engineering and communication through statistics, and can be divided into [109]: circular, Toeplitz's, Hankel's, Vandermonde's, Cauchy, Frobenius's, Sylvester's, Bzout, Loewner's, and Pick's matrix. What is generally known is that these matrices have a very characteristic structure, and hence the title of the structural matrix. This specific structure contained in their primary refers to a feature that is used to store these matrices enough to remember much smaller elements of the total number of elements. For example, in some cases, it is sufficient to remember only one series consisting of O(n) of elements where n is the dimension of the matrix (the Toeplitz-Hank's, circular, Vandermonde's) or perhaps two sets as is the case with the Cauchy matrix. The following definition is administered in this regard.

Definition 1.1.1. A dense matrix of order n–n matrix, if we call structural elements this matrix depend only on the O(n) parameters. As the following characteristics should be noted that in the study of these matrices is very important achieving reduction in the calculation, especially in the multiplication of these matrices between themselves, as well as in their vector multiplication. Using a discrete Fourier transform and cosine transform, very fast algorithms were built for matrix multiplication of these vectors. These algorithms will be dealt with in more detail in the second chapter of this thesis. Studying relations (links), and calculating the similarity between polynomial and rational functions for calculating the structural matrices came to the improvements on the calculation of both. Specifically, the operation multiplication, division, and other interpolation calculating the polynomial solutions which are obtained mainly vases and structural matrix, or are closely related to them. In some cases, two resulting algorithms are actually identical. This means that as an outlet and as means getting the same results, but they have only different statements and representations. Some are given in terms of the polynomial and the second in terms of the structural stencil. This same approach for solving various problems of finding different gives must be securely improvement, expansion and generalization of the problems posed. In many cases when solving given problem algorithms for calculating the polynomials become quick, while on the other side algorithms for calculating the structural matrices show bigger numerical stability. Like conclusion obtained by two related objectives in the design process algorithms:

Lemma 1.0.1. *1. Make an effective algorithm for polynomials and rational calculation; and*

2. Make an effective algorithm for calculating the structural matrix.

Calculating the structural matrix, if properly performed, is done in a very shorter time and takes up much less storage space than ordinary calculations on matrices. The main problem that arises here is the need to

perform the calculations properly. In order to calculations performed properly used and investigating access site rank matrix (Displacement rank approach). The general idea is to describe the deployment matrix M as image L (M) site linear operator L applied to the matrix M. What is important is that the matrix L (M) remains small while the size of the matrix M increases. Later, we have to Matrix M may present using operators L and matrix L(M). Based on all of the above characteristics and which can be in a wide literature found for this class of matrices emphasize the following.

Lemma 1.0.2. *1. may be represented by a small number of using parameters (elements);*

2. *their multiplication with a vector can not occur quickly, almost in linear time;*
3. *have a close algorithmic correlation (relationship) with calculation of the polynomial and rational functions, particularly with their multiplication, division, and interpolation;*
4. *in their natural connection with the deployed linear operator L (displacement operator) and in a simple manner may be presented using operators L and matrix L(M), which have small rank.*

These classes matrix can be expanded and generalized to other classes of structural matrix as the f-circular Toeplitz's matrices, strip Toeplitz's matrices (banded Toeplitz matrices) block Toeplitz's matrices, Toeplitz plus Hankel's matrix, block matrix with Toeplitz, Hankel or Vandermone's blocks, as well as many others. Structural matrix are extremely important and have great application in science, technology and in communications. Many important problems in applied mathematics and technology can be reduced to some matrix problem. Moreover, to meet the needs of different applications of these problems usually special structure on corresponding matrix are introduced in a way that the elements of these matrices can be described with same rules (formulas). The classic examples include Toeplitz's, Hankel's, Vandermonde, Cauchy, Pick-ups, Bzout's and as well as other structural matrix. Using these, some more general structure usually leads to elegant mathematical solution of the problem as well as to the construction, practical, and efficient algorithms for a multitude of different problems that can occur in engineering and science. For many years, much attention has been given to the study of the structural matrix and from various fields such as mathematics, computing and engineering. A considerable progress in all these areas has been made, especially in terms of calculating the structural matrices and an examination of their numerical characteristics. In this respect, the structural matrix, for natural way, imposes the position of the bridge (connection) between some applications in science and technology—on the one hand, theoretical mathematics and on the other hand, calculations and numerical characteristics. Many presented papers on this subject, such as [3, 25], confirm the fact that the different methods in engineering, mathematics, and numerical analysis complement each other and they all contribute to the development of unified theory of structural matrix.

Analysis of theoretical and numerical characterization matrix structure is of fundamental importance to design efficient algorithms for calculating the matrices and polynomials. Some general algorithms for solving many problems, large size, they become useless because of its complexity. Therefore, specially designed algorithms that rely on some special features, common to a class of problems remain efficient and in the case of large dimensions of the problem. It actually means that the wealth of features of many classes of structural matrix allows researchers to construct and analyze the fast algorithms for solving the problems that are associated with these classes of matrices. Theoretical achievements were used in order to obtain efficient numerical algorithms for solving various problems concerning the structural matrix. In addition, some other problems, which are actually very far from the structural matrix, have been reformulated in terms of the structural matrix and effectively solved by means of appropriate algorithms. Results obtained in this way are used for solving many other problems in different areas.

One of the most significant and the most studied classes of matrices is a class of Toeplitz' matrices. These matrices appear and are used in image processing, signal processing, digital filtering, operations theory, computer algebra, as well as in the field of finding solutions certain of difference and differential equations numerically. Another class of structural matrices are a little less technology applications, but also are very important for studying how a theoretical point of view and from the point of application. Frobenius's, Hankel's, Sylvester's, and Bezoutians's matrices appearing in the theory of control (control theory), the theory of stability and calculation of the polynomials. These matrices possess many important characteristics and are subject to many scientific research. Vandernomde's, Cauchy, Loewner's and Pick's matrices are mainly applied in the context of interpolation problems. Tridiagonal matrices and some other more general, band matrices and their inverses belonging to the semi-separable matrices are very closely related to numerical analysis. Expansion of the matrix to some more general class and design efficient algorithms in previous years receive a lot of attention scientific community. Also, some more dimensional problems lead to the emergence of a matrix that can be represented using structural block matrix, the matrix whose blocks have characteristic structure.

Matrix algebra is another field in which the structural matrix play an important role. Circular dies and the Hartley algebra are examples of trigonometric algebra that have very rich algebraic and numerical properties. Their connections with fast discrete transformations, such as Fast Fourier Transform (FFT), make them a useful tool for the design of effective matrix calculation code algorithms.

Under the matrix with symbolic elements, we mean a matrix whose elements are from the relevant commutative ring. Many applications include the problem of inverting a matrix with symbolic elements [1] or similar problems, for example, creating the matrix minora. Factorization polynomial matrices A, i.e., matrix whose elements are polynomials, are often used to calculate the inverse and the generalized inverse matrix A, [36, 51]. When the matrix A nonsingular, Gauss's algorithm is an efficient way for determining the inverse A^{-1}. Some polynomial algorithms in computer algebra were observed in the work [85]. Software packages such as `MATHEMATICA` [86], `MAPLE` and `MATLAB`, which include symbolic calculations, contain built-in functions for matrix factorization. However, due to difficulties with the simplification, it is not practical to implement the matrix factorization on polynomial or rational matrix in procedural programming languages. Also, the decomposition of the polynomial (rational) matrix—or matrix elements are polynomials (rational functions)—is quite different and more complicated (see [36, 51]). This idea is also applied to the integer-preserving elimination used for solving systems of equations for the case of univariate polynomials or integers. Several recent advancement in this field, reduce the complexity of most problems with polynomial matrices in the order of magnitude of complexity of matrix multiplication, which is so far the biggest achievement with the complexity [18, 28, 38, 70].

Calculation of generalized inverse of a polynomial matrix $A(x)$ is an important problem of numerical analysis, which often occurs in inverse systems [40] and system solutions of Diophantine equations [30]. One of the algorithms for calculating the generalized inverse matrix polynomial $A(x)$ Karampetak is introduced in Ref. [30]. Based on the idea that the polynomial matrix can be represented as a list of constant matrix, several new algorithms to calculate different types of generalized inverse is performed in the works [75, 76]. One of the difficulties in these algorithms is their dependence on the degree of the matrix $A(x)$. In this sense, we have introduced modifications known algorithms based on the factorization of the full-rank of rational matrix, like improvement of calculation time and stability.

One possible approach is to apply QDR and LDL^* decomposition of the full-rank Hermitian polynomial matrix A (methods obtained in Ref. [72]) and the representation of generalized inverse $A^{(2)}_{T,S}$ from Ref. [53]. Methods introduced in [53] are derived on the basis of FG full-rank decomposition of matrix A. Thus, the expansion of this algorithm can be performed that is applicable for the calculation of the broad class $A^{(2)}_{T,S}$ inverses, not just Moore–Penrose's inverse. Moreover, developed the expansion of this algorithm on a set of polynomial matrix of one variable. Instead of the FG matrix decomposition A we used QDR and LDL^* decomposition full-rank appropriate selected matrix M. Observe that the choice of LDL^* factorization is crucial to eliminate appearances of square-root entries.

1.1 Basic Definitions and Properties

Generalized (general) inverse matrix represents a generalization of the concept of ordinary matrix inversion. If A is a regular square matrix, i.e., If $detA6 = 0$, then there is a unique matrix X such that $AX = XA = I$, where I is a unit matrix. In this case, X is the inverse matrix A and matrix marked with $A - 1$. If A is a singular matrix (or rectangular matrix), then the matrix X with these properties does not exist. In these cases, it is useful to determine the kind of "inverse" matrix A, i.e., matrix keep as much more properties as inverse matrix. This led to the concept of generalized inverse matrix A. General inverse matrix includes matrix X, which is in a sense an associated matrix. And so, the following is valid:

Lemma 1.1.1. *1. general inverse matrix exists for the class which is wider than the class of regular matrices (in some cases of an arbitrary matrix A);*

2. there are some properties of ordinary inverse;

3. comes down to the ordinary inverse when A nonsingular square matrix.

The idea of a generalized inverses is still implicitly contained in the works of C. F. Gauss from 1809 and in connection with introducing principle of least-squares method inconsistent system. After that I. Fredholm in 1903 defined linear integral operator, which is not invertible in the ordinary sense, solve the integral equation in the cases when the inverse operator exists. It turned out that generally, inverse operator is not unique. W. A. Hurwitz in 1912, using resolutions of the term 'pseudo' described whole class of such operators. Generalized inverses of differential operators are implicitly contained in Hilbert considering the generalized Green's functions in 1904 and later were studied. Other authors, for example. W. T. Reid in 1931, and so on. H. E. Moore in 1920 first defined and studied the unique generalized inverse of an arbitrary matrix, calling it a "generalized reciprocity matrix". It is possible that the first results of Moore came in 1906, although the first results were published only in 1920. However, his rad little was known to a broader audience, possibly because of the specific terminology and labels. Only in 1955 the work of R. Penrose has aroused genuine interest in the study of this issue. Penrose proved that Moore actually the inverse matrix equation system solution and therefore the inverse today called Moore–Penrose inverse. Penrose is also pointed to the role of the generalized inverse solving systems of linear equations. The theory and application of methods for the calculation of generalized inverses were developing very quickly in the last 50 years, with publishing of a number of scientific papers and several monographs.

The set of all complex $m \times n$ matrices is marked with $\mathbb{C}^{m\times n}$, while the $\mathbb{C}_r^{m\times n} = \{X \in \mathbb{C}^{m\times n} \mid rank(X) = r\}$ a subset of it which consists of a complex matrix of rank k. Let $\mathbb{C}(x)$ (respectively $\mathbb{R}(x)$) means a set of rational functions with complex (respectively real) coefficients. Then the set of all $m \times n$ matrix with elements in $\mathbb{C}(x)$ (respectively $\mathbb{R}(x)$) indicates the $\mathbb{C}(x)^{m\times n}$ (respectively $\mathbb{R}(x)^{m\times n}$).

With I_r and I, we denote identity matrix of order r, or unit matrix appropriate order, respectively. Whereas $\mathbf{O}$ denotes the zero matrix of appropriate order of magnitude.

For an arbitrary matrix $A \in \mathbb{C}(x)^{m\times n}$ consider the following matrix equation for the unknown X, where $*$ denotes conjugate transpose of the matrix:

$$(1)\quad AXA = A \quad (2)\quad XAX = X$$

$$(3)\quad (AX)^* = AX \quad (4)\quad (XA)^* = XA.$$

In the case $m = n$, observe the following matrix equation

$$(5)\quad AX = XA \qquad (1^k)\quad A^{k+1}X = A^k.$$

Now, consider the set $\mathscr{S} \subset \{1,2,3,4\}$. Then a set of matrices that satisfy equations with indices from the set $\mathscr{S}$ is denoted by $A\{\mathscr{S}\}$. An arbitrary matrix $A\{\mathscr{S}\}$ is called $\mathscr{S}$-generalized inverse of the matrix A. Matrix

$X = A^\dagger$ the Moore–Penrose's inverse matrix A if it satisfies the Eqs. (1)–(4). Group inverse $A^\#$ is a unique $\{1,2,5\}$ inverse of the matrix A, and exists if and only if it is valid that

$$\mathrm{ind}(A) = \min_k\{k : rank(A^{k+1}) = rank(A^k)\} = 1.$$

Rank of generalized inverse $A^\dagger$ is often used in the calculation, and it is necessary to consider a subset $A\{i,j,k\}_s$ of the set $A\{i,j,k\}$, which consists of $\{i,j,k\}$-inverses of rank s (see [3]).

By I, we denote the identity matrix of an appropriate order, and $\mathbf{O}$ denotes an appropriate zero matrix. In this book, we denote rank matrix $A \in \mathbb{C}(x)^{m\times n}$ sa $\mathscr{R}(A)$, and kernel of matrix A with $\mathscr{N}(A)$. Also, with $nrang(A)$ we denote the normal rank of matrix A rank over the set $\mathbb{C}(x)$) and with $\mathbb{C}(x)_r^{m\times n}$ we denote set of rational matrix of $\mathbb{C}(x)^{m\times n}$ normal rank r. Similarly, $rank(A)$ denotes rank of constant matrix A and $\mathbb{C}_r^{m\times n}$ denotes a set of complex matrix $\mathbb{C}^{m\times n}$ with rank r. Here are state from Ref. [3, 80] which defines $A_{T,S}^{(2)}$ generalized inverse.

Proposition 1.1.2. [3, 80] *If $A \in \mathbb{C}_r^{m\times n}$, T is subspace of space $\mathbb{C}^n$ with dimensions $t \le r$ and S is subspace of $\mathbb{C}^m$ with dimensions $m-t$, then matrix A has $\{2\}$-inverse X such that $\mathscr{R}(X) = T$ i $\mathscr{N}(X) = S$ if and only if $AT \oplus S = \mathbb{C}^m$, and in that case X is unique and we denote with $A_{T,S}^{(2)}$.*

Probably the most studied class of generalized inverse is the inverse of external classes with defined images and kernels. Generalized inverses, as Moore–Penrose $A^\dagger$, weight Moore–Penrose $A_{M,N}^\dagger$, Drazin A^D and group inverse $A^\#$, and Bott-Duffin inverse $A_{(L)}^{(-1)}$ and generalized Bott-Duffin inverse $A_{(L)}^{(\dagger)}$, can unite in this way and represent as a generalized inverses $A_{T,S}^{(2)}$ (in what are considered appropriate matrix T i S). For given matrix

$A \in \mathbb{C}^{m\times n}(x)$ = the next representation is valid (for more details see [3, 80]):

$$A^{\dagger} = A^{(2)}_{\mathscr{R}(A^*),\mathscr{N}(A^*)}, \quad A^{\dagger}_{M,N} = A^{(2)}_{\mathscr{R}(A^{\sharp}),\mathscr{N}(A^{\sharp})}, \tag{1.1.1}$$

where M, N are positive definite matrix of appropriate dimensions, $A^{\sharp} = N^{-1}A^*M$.

Square matrix is given $A \in \mathbb{C}^{n\times n}(x)$ and Drazin inverse *i* group inverse are given with the following expressions (see [3, 80]):

$$A^{D} = A^{(2)}_{\mathscr{R}(A^k),\mathscr{N}(A^k)}, \quad A^{\#} = A^{(2)}_{\mathscr{R}(A),\mathscr{N}(A)}, \tag{1.1.2}$$

where is $k = \operatorname{ind}(A)$.

If L is subspace of $\mathbb{C}^n$ and if A L-positive semi-definite matrix which satisfies conditions $AL \oplus L^{\perp} = \mathbb{C}^n$, $S = \mathscr{R}(P_L A)$, then, the following identity are satisfied, [6, 80]:

$$A^{(-1)}_{(L)} = A^{(2)}_{L,L^{\perp}}, \quad A^{(\dagger)}_{(L)} = A^{(2)}_{S,S^{\perp}}. \tag{1.1.3}$$

Effective methods for calculating weight divisions, Moore–Penrose's inverse is made in Ref. [45].

In this book the direct method for the calculation of generalized inverse matrix is studied. By direct methods, each element is calculated generalized inverse without explicit iterative numerical improvements. These are some of the famous group of methods of representation of generalized inverse:

1. generalized inverse calculation using the full-rank factorization;
2. the method based on matrix factorizations;
3. the method arising from the decomposition of block;
4. the method of alterations (*partitioning method*);

5. determinant representation;

6. the method is based on the generalization of Frame's results; and

7. the method is based on breaking the matrix (*matrix splitting*).

1.2 Representation of Generalized Inverses Using Full-Rank Decomposition

It is generally known that the following result holds each matrix decomposition of the full-rank.

Lemma 1.2.1. *Each matrix $A \in \mathbb{C}_r^{m\times n}$ can be presented in the form of products $A = PQ$ two full-rank matrix $P \in \mathbb{C}_r^{m\times r}$ i $Q \in \mathbb{C}_r^{r\times n}$.*

This decomposition is not unique. However, if one of the matrix P or Q specified, then the other matrix uniquely determined. Full-ranking factorization can be used to calculate various classes generalized inverse. For the full-rank factorization are based very useful factorization of generalized inverse.

Here are some results of the general presentation of generalized inverses, in terms of full-rank factorization and adequately selected matrix. In the next lemma, we will show the equivalence between the two representations $\{1,2\}$-inverse, which is proved in Ref. [48].

Lemma 1.2.2. *Let $A = PQ$ be full-rank factorization of $A \in \mathbb{C}_r^{m\times n}$. Let U, V be matrix which papers belonging to the matrix type $m \times m$ and $n \times n$ respectively. Also, let matrix W_1 i W_2 belongs to sets of the matrix type*

$n \times r$ i $r \times m$ respectively. Then the sets

$$S_1 = \{VQ^*(P^*UAVQ^*)^{-1}P^*U : U \in \mathbb{C}^{m\times m}, V \in \mathbb{C}^{n\times n},$$
$$\text{rank}(P^*UAVQ^*) = r\}$$
$$S_2 = \{W_1(W_2AW_1)^{-1}W_2 : W_1 \in \mathbb{C}^{n\times r}, W_2 \in \mathbb{C}^{r\times n},$$
$$\text{rank}(W_2AW_1) = r\}$$

satisfies $S_1 = S_2 = A\{1,2\}$.

Proof. Using the known result [48], it is easy to conclude that $X \in \{1,2\}$, if and only if it can be written in the form

$$X = VQ^*(QVQ^*)^{-1}(P^*UP)^{-1}P^*U = VQ^*(P^*UAVQ^*)^{-1}P^*U,$$
$$U \in \mathbb{C}^{m\times m}, V \in \mathbb{C}^{n\times n}, \text{rank}(P^*UAVQ^*) = r.$$

This implies $S_1 = A\{1,2\}$.

Further, in order to complete the proof, show that $S_1 = S_2$. Suppose that $X \in S_1$ arbitrarily. Then exists matrix $U \in \mathbb{C}^{m\times m}$ and $V \in \mathbb{C}^{n\times n}$ such that $rank(P^*UAVQ^*) = r$ I X can be presented in term of $X = VQ^*(P^*UAVQ^*)^{-1}P^*U$. After changing $VQ^* = W_1 \in \mathbb{C}^{n\times r}, P^*U = W_2 \in \mathbb{C}^{r\times m}$, we directly get $X \in S_2$. On the other hand, suppose that $X \in S_2$. Then exists $W_1 \in \mathbb{C}^{n\times r}$ i $W_2 \in \mathbb{C}^{r\times m}$, such that $rank(W_2AW_1) = r$ i $X = W_1(W_2AW_1)^{-1}W_2$. Matrix equations $W_1 = VQ^*$ i $W_2 = P^*U$ are consistent. Really, equation $W_1 = VQ^*$ is consistent if and only if $Q^{*^{(1)}}$ such that $W_1Q^{*^{(1)}}Q^* = W_1$. For example, we can take an arbitrary left inverse of Q^*. The consistency of the matrix equality $W_2 = P^*U$ is proved in a similar way. Thus, for some matrix $U \in \mathbb{C}^{m\times m}$ i $V \in \mathbb{C}^{n\times n}$, randomly selected matrix $X \in S_2$ can be written in the form $X = VQ^*(P^*UAVQ^*)^{-1}P^*U$, $rank(P^*UAVQ^*) = r$, which implies $X \in S_1$. □

In the following [62] introduced a general presentation class {2}-inverse.

Theorem 1.2.3. *Set of* $\{2\}$*-inverse of the given matrix* $A \in \mathbb{C}_r^{m\times n}$ *specifies the following equality:*

$$A\{2\} = \{W_1(W_2AW_1)^{-1}W_2, W_1 \in \mathbb{C}^{n\times t}, W_2 \in \mathbb{C}^{t\times m},$$

$$rank(W_2AW_1) = t, t = 1, \ldots, r\}$$

Proof. Let X be an arbitrary $\{2\}$-inverse of A. Applying the theorem 3.4.1 from Ref. [48], it can be presented in a generalized form $X = C(DAC)^{(1,2)}D$, where $C \in \mathbb{C}^{n\times p}$ and $D \in \mathbb{C}^{q\times m}$ are arbitrary. Also, it is known da $rank(X) = rank(DAC)$ [48] [Theorem 3.4.1]. WE suppose that $rank(DAC) = t \leq min\{p,q\} \leq r$. Applying Lemma 1.1.2 we obtain generalized presentation for X:

$$X = CJ(HDACJ)^{-1}HD, J \in \mathbb{C}^{p\times t}, H \in \mathbb{C}^{t\times q}, \text{rank}(HDACJ) = t.$$

After replacing $CJ = W_1$, $HD = W_2$, it is easy to note that the

$$X \in \{W_1(W_2AW_1)^{-1}W_2, W_1 \in \mathbb{C}^{n\times t}, W_2 \in \mathbb{C}^{t\times m},$$

$$\text{rank}(W_2AW_1) = t,\ t = 1, \ldots, r\}.$$

Conversely, for an arbitrary X given with

$$X \in \{W_1(W_2AW_1)^{-1}W_2, W_1 \in \mathbb{C}^{n\times t}, W_2 \in \mathbb{C}^{t\times m},$$

$$\text{rank}(W_2AW_1) = t,\ t = 1, \ldots, r\}.$$

it is easy to show that satisfies the equality $XAX = X$. □

In the next theorem from Ref. [62], a general presentation Drazins square inverse matrix A is introduced in terms of full-rank factorization invariant degree A^l, $l \geq ind(A)$ of matrix A.

Theorem 1.2.4. *Let $A \in \mathbb{C}_r^{n\times n}$, $k = ind(A)$ and $l \geq k$ is an arbitrary integer. If $A^l = PQ$ is full-rank factorization of matrix A^l, then Drazin inverse of A has the following presentation:*

$$A^D = P(QAP)^{-1}Q.$$

Proof. Let $T \in \mathbb{C}^{n\times n}$ be nonsingular matrix such that

$$A = T\begin{bmatrix} C & \mathbb{O} \\ \mathbb{O} & N \end{bmatrix} T^{-1}$$

kernel-nullpotent separation of A, where C is nonsingular and N nullpotent matrix with index k. It is easy to notice

$$A^l = T\begin{bmatrix} C^l & \mathbb{O} \\ \mathbb{O} & \mathbb{O} \end{bmatrix} T^{-1}$$

for each $l \geq ind(A)$. This implies the following full-rank decomposition $A^l = PQ$:

$$P = T\begin{bmatrix} C^l \\ \mathbb{O} \end{bmatrix}, Q = \begin{bmatrix} I_r, & \mathbb{O} \end{bmatrix} T^{-1}.$$

Using this fact and know the canonical form of representation for A^D [5],we get

$$A^D = T\begin{bmatrix} C^{-1} & \mathbb{O} \\ \mathbb{O} & \mathbb{O} \end{bmatrix} T^{-1} = T\begin{bmatrix} C^{-1} \\ \mathbb{O} \end{bmatrix}$$

$$\begin{bmatrix} I_r, & \mathbb{O} \end{bmatrix} T^{-1} = PC^{-(l+1)}Q.$$

We complete proof with

$$C^{l+1} = \begin{bmatrix} C, & \mathbb{O} \end{bmatrix}\begin{bmatrix} C^l \\ \mathbb{O} \end{bmatrix} = \begin{bmatrix} I_r, & \mathbb{O} \end{bmatrix}$$

$$T^{-1}T\begin{bmatrix} C & \mathbb{O} \\ \mathbb{O} & N \end{bmatrix} T^{-1}T\begin{bmatrix} C^l \\ \mathbb{O} \end{bmatrix} = QAP.$$

□

Remark 1.2.1. Representation of the Drazin inverse introduced in 1.2.4 is a natural generalization of the following characterization of inverse groups introduced in [9]:

If $A = PQ$ is full-rank factorization of A, then exists $A^{\#}$ if and only if the matrix QP is invertible, and it is valid

$$A^{\#} = P(QP)^{-2}Q = P(QAP)^{-1}Q.$$

Full-rank matrix factorization can be determined in several ways. Some methods of full-rank matrix factorization can take a variety of matrix decomposition, such as *LU* decomposition, *QR* factorization, *SVD Singular Value-based Decomposition*, and others. Full-rank factorization can be obtained using the any of the blocks of the matrix representation. These methods will be studied in the next section.

1.3 LU Decomposition

This book observers that General LU decomposition, and its variants: Cholesky, or LL * decomposition and LDL * decomposition. The following assertion is well-known from the literature.

Proposition 1.3.1. *Each matrix $A \in \mathbb{C}_r^{m\times n}$ can be transformed into a form $A = LU$, where L is a lower triangular matrix with unit the main diagonal, a U is an upper triangular matrix.*

Proof. *LU* factorization can be derived from the Gauss elimination method by the complete selection of the main elements. Matrix $A = A_0 \in \mathbb{C}_r^{m\times n}$ is transforming to the matrices $A_1, \dots, A_r$, respectively. Matrix A_k, $0 \le k \le r$ is $m \times n$ matrix of the form

$$A_k = \left(a_{ij}^{(k)} \right) = \begin{bmatrix} U_k & V_k \\ \mathbb{O} & W_k \end{bmatrix},$$

where U_k is the triangular matrix of dimension $k \times k$. Denote (i, j)-element in block W_k with w_{ij}. If $w_{\alpha\beta}$ is the biggest element modulo in block W_k is performed by replacing α row with $k+1$ row and β row with $k+1$ column of matrix A_k. Execution involves the replacement of the types of identity $m \times m$ matrix obtained from permutation matrix $E_k = P(k+1,)$, while replacing analog column unitary $n \times n$ matrix obtained permutation matrix $F_k = P(k+1, \beta)$. Assuming $a_{ij}^{(1)} = a_{ij}, \quad 1 \le i \le m,\ 1 \le j \le n$, Gaussian eliminating of matrix A_k is the next transformation:

$$v_{ik} = \frac{a_{ik}^{(k)}}{a_{kk}^{(k)}},$$

$$a_{ij}^{(k+1)} = a_{ij}^{(k)} - v_{ik} a_{kj}^{(k)}, \begin{pmatrix} i = k+1, \ldots, m \\ j = k+1, \ldots, n \end{pmatrix}, k = 1, \ldots, r.$$

As a result of this transformation matrix is obtained by

$$A_{k+1} = \begin{bmatrix} U_{k+1} & V_{k+1} \\ \mathbb{O} & W_{k+1} \end{bmatrix}.$$

After r steps yields matrix

$$A_k = \begin{bmatrix} U_r & V_r \\ \mathbb{O} & W_r \end{bmatrix} = \begin{bmatrix} a_{11}^{(1)} & a_{12}^{(1)} & \ldots & a_{1r}^{(1)} & \ldots & a_{1n}^{(1)} \\ 0 & a_{22}^{(2)} & \ldots & a_{2r}^{(2)} & \ldots & a_{2n}^{(2)} \\ \ldots & \ldots & \ldots & \ldots & \ldots & \ldots \\ 0 & 0 & \ldots & a_{rr}^{(r)} & \ldots & a_{rn}^{(r)} \end{bmatrix}$$

For matrix U can be taken first r matrix type A_R.

The matrix L is defined by

$$L = \begin{bmatrix} 1 & 0 & \dots & 0 & 0 \\ v_{21} & 1 & \dots & 0 & 0 \\ \dots & \dots & \dots & \dots & \dots \\ v_{r1} & v_{r1} & \dots & v_{r,r-1} & 1 \\ \dots & \dots & \dots & \dots & \dots \\ v_{m1} & v_{m2} & \dots & v_{m,r-1} & v_{m,r} \end{bmatrix}. \quad \square$$

At the end of the transformation is given by:

$$EAF = LU \Longleftrightarrow A = E^* LUF^*,$$

where E and F are permutation matrix defined as product of elementary matrices:

$$E = E_1 \cdots E_r, \quad F = F_1 \cdots F_r.$$

Proposition 1.3.2. *If $A = LU$ presents LU factorization of matrix A, then*

$$A^\dagger = U^\dagger L^\dagger = U^*(UU^*)^{-1}(L^*L)^{-1}L^*.$$

Below is a description of the implementation of the *LU* decomposition.

Conjugate and transposed matrix is formed by the following features:

```
Hermit[a_]:= Conjugate[Transpose[a]
```

Rank of the matrix is calculated as the number of nonzero type of matrix that is obtained. Application of elementary transformation of type can be accomplished using the tool *RowReduce*:

```
rank[a_]:=
    Block[{b=a,m,n,r=0,i,nula},
    {m,n}=Dimensions[b];
    b=RowReduce[b];
    nula=Table[0,{n}];
    Do[If[b[[i]]!=nula,r++],{i,m}];
    Return[r]
```

```
];
```

Now *LU* factorization of the matrix can be generated as follows:

```
LU[a_]:=
  Block[{b=a,e,f,n,m,l,r,ran,i,k,z,c,p={},d},
  {m,n}=Dimensions[b];
  ran=rank[b];
  e=IdentityMatrix[m];
  f=IdentityMatrix[n];
  l=Table[0,{c,m},{d,ran}];
  r=Table[0,{c,ran},{d,n}];
  Do[b1=zameni[b,e,f,n,m,k,k];
     b=b1[[1]];e=b1[[2]];
     f=b1[[3]];,{k,1,ran}];
  Do[
  Do[
  If[b[[k,k]]!=0,
  l[[i,k]]=b[[i,k]]/b[[k,k]];
  Do[b[[i,j]]-=l[[i,k]]*b[[k,j]],{j,k+1,n}],
  Print[Singular matrix!!!"];Exit];
  {i,k+1,m}],{k,1,ran}];
  Do[l[[c,c]]=1,{c,ran}];
  Do[If[c<=d,r[[c,d]]=b[[c,d]]],{c,ran},{d,n}];
  Print[L=,l];
  Print[U=,r];
  Print[E=,b1[[2]]];
  Print[F=,b1[[3]]];
  Print[L.U=,l.r];
  Print[A=E*.L.U.F*=,Inverse[e].(l.r).Inverse[f]];
  Return[{Inverse[e].l,r.Inverse[f]}];
  ];
```

```
replace[a_,e_,f_,n_,m_,p1_,p2_]:=
    Block[{b=a,e1=e,f1=f,s,i,j,k,l},
    s=Abs[b[[p1,p2]]];k=p1;l:=p2;
    Do[
    Do[
    If[s<Abs[b[[i,j]]],s=Abs[b[[i,j]]];
    k=i;l=j],{j,p2,n}],{i,p1,m}];
    b=ChRows[b,p1,k];e1=ChRows[e1,p1,k];
    b=ChColumns[b,p2,l];f1=ChColumns[f1,p2,l];
    Return[{b,e1,f1}];
    ];
```

Example 1.3.1. For matrix $A = \begin{bmatrix} 1 & 3 & 6 & 9 \\ 1 & 2 & 3 & 4 \\ 0 & -2 & 0 & -2 \\ -1 & 0 & -1 & -3 \end{bmatrix}$ we get LU factorization:

$$L = \begin{bmatrix} 1 & 0 & 0 & 0 \\ \frac{4}{9} & 1 & 0 & 0 \\ -\frac{2}{9} & 4 & 1 & 0 \\ -\frac{1}{3} & 3 & \frac{1}{4} & 1 \end{bmatrix}, \quad U = \begin{bmatrix} 1 & 3 & 6 & 9 \\ \frac{5}{9} & \frac{2}{3} & \frac{1}{3} & 0 \\ -2 & -4 & 0 & 0 \\ -\frac{11}{6} & 0 & 0 & 0 \end{bmatrix}.$$

The program calls on the expression form

$\{e, L, U, f\} = lu[\{\{1,3,6,9\},\{1,2,3,4\},\{0,-2,0,-2\},\{-1,0,-1,-3\}\}]$

```
lump[a_]:=
    Block[{b=a,l,u},
    {l,u}=lu[b];
    Hermit[u].Inverse[u.Hermit[u]].Inverse[Hermit[l].
l].Hermit[l]
    ]; MatrixQ[a]
```

Implemented generalized full-rank factorization is applied, expressed in terms of blocks contained in various matrix factorizations, such as LU-factorization, QR-factorization.

Theorem 1.3.3. *Let $EAF = LU$ be LU-factorization of $A \in \mathbb{C}_r^{n\times n}$, where A $m \times r$ is matrix with zeros on the main diagonal and U is $r \times n$ upper trapezoidal matrix.Then*

$$X \in A\{2\} \Leftrightarrow X = FY(ZLUY)^{-1}ZE, rank(ZLUY) = t,$$

where $Y \in \mathbb{C}^{n\times t}, Z \in \mathbb{C}^{t\times m}, t \in \{1,\ldots,r\}$;
For $Y \in \mathbb{C}^{n\times r}$ i $Z \in \mathbb{C}^{r\times m}$ we get

$$X \in A\{1,2\} \Leftrightarrow X = FY(ZLUY)^{-1}ZE, rank(ZLUY) = r;$$

$$X \in A\{1,2,3\} \Leftrightarrow X = FY(L^*LUY)^{-1}L^*E, rank(L^*LUY) = r;$$

$$X \in A\{1,2,4\} \Leftrightarrow X = FU^*(ZLUU^*)^{-1}ZE, rank(ZLUU^*) = r;$$

$$A^\dagger = FU^*(L^*LUU^*)^{-1}L^*E;$$

$A^\#$ exists if and only if UF^*E^*L is invertible, and $A^\# = E^*L(UF^*E^*L)^{-2}UF^*$.

LU factorization is the program package `MATHEMATICA` implemented using standard function `LUDecomposition`. The term `LUDecomposition [m]` produces *LU* decomposition numerical *M* matrix. The result is a list of $\{l,u\}$, where l lower triangular matrix, and u is an upper triangular matrix. The starting matrix m is equal to `l.u`.

The calculation of Moore–Penrose using inverse *LU* decomposition can implement the following functions:

```
LUMP[a_]:=
   Block[{b=a,l,u},
   {l,u}=LUDecomposition[b];
```

```
    Return[Hermit[u].Inverse[Hermit[l].a.
Hermit[u]].Hermit[l]];
    ]
```

Cholesky factorization involves decomposition of symmetric positive definite matrix A in form $A = LL^*$, where L is a lower triangular matrix with strictly positive diagonal elements. Obviously, the elements of the matrix L can contain square roots, which causes difficulties in symbolic calculations. For this reason "square-root-free" Cholesky decomposition in the form of $A = LDL^*$ is often used. An arbitrary rational Hermitian matrix A can be represented as a product matrix $A = LDL^*$, where L is lower triangular, and D diagonal matrix. This form of the bridges uses the elements containing square roots. It is easy to show that, in the case of positive definite matrix A, the diagonal elements of the D must be non-negative.

Let a Hermitian polynomial matrix $A(x) \in \mathbb{C}(x)^{n\times n}$ be given with elements containing variable x:

$$A(x) = A_0 + A_1x + \cdots + A_qx^q = \sum_{i=0}^{q} A_ix^i \tag{1.3.1}$$

wherein A_i, $i = 0, \ldots, q$ are constant matrices with dimension $n \times n$. Alternative form of Cholesky factorization symmetric matrix polynomial $A(x)$ generates an additional diagonal matrix $D(x)$, in order to avoid the elements with square roots (see Ref. [19]). Thus, LDL^* factorization of matrix $A(x)$ is square root free Cholesky decomposition in the following form:

$$A(x) = L(x)D(x)L^*(x),$$

where $L(x)$ is lower triangular, a $D(x)$ diagonal matrix,

$$L(x) = \begin{bmatrix} 1 & 0 & \cdots & 0 \\ l_{21}(x) & 1 & \cdots & 0 \\ \vdots & \vdots & \ddots & \vdots \\ l_{n1}(x) & l_{n2}(x) & \cdots & 1 \end{bmatrix}, \tag{1.3.2}$$

$$D(x) = \begin{bmatrix} d_1(x) & 0 & \cdots & 0 \\ 0 & d_2(x) & \cdots & 0 \\ \vdots & \vdots & \ddots & \vdots \\ 0 & 0 & \cdots & d_n(x) \end{bmatrix} \tag{1.3.3}$$

wherein l_{ij}, d_j, $1 \le j < i \le n$ are rational functions. Thereby, $L^*(x)$ denotes conjugate transpose matrices $L(x)$.

Replacing LL^* decomposition with LDL^* factorization in order to avoid calculating square root elements there is well-known method of numerical analysis, described by Golub and Van Loan [19]. Computation of square-root entries in the Cholesky decomposition can be avoided by using the LDL^* factorization. This technique is of essential importance for polynomial and rational elements of the matrix L i D. If, for instance, we have $\sqrt{\sum_{i=0}^{q} A_i x^i}$ symbolic, where A_i are, $i = 0,\ldots,q$ constant $n \times n$ matrices, it can be very difficult. Therefore, it is the motivation modifying algorithms with direct Cholesky factorization polynomial matrix $A(x)$. It is evident that this form is appropriate for manipulation with polynomial matrix.

Accordingly, the full-rank decomposition can be used for the symbolic generalized inverse calculation using a number of techniques introduced in the works, such as [65, 53]. The implementation of these algorithms solve the problem of symbolic calculations generalized inverse in procedural programming languages.

CHAPTER 2

COMPUTING GENERALIZED INVERSES OF MATRICES

This chapter deals with the definitions, basic properties, and some methods for computing generalized inverses of constant matrices. For all of these methods, algorithms are given based on such methods. The time complexity of introduced algorithms is also given.

Most of the results are known and taken from literature, except the last section which describes a new method for rapid computation of Moore–Penrose (MP) and some $\{i, j, \ldots, k\}$ inverses.

2.1 Preliminaries and Notation

In this section, we will define several classes of generalized inverses and study their basic properties. For the ease of reference, we collect here facts, definitions, and notations that are used throughout this and next chapter.

Let $\mathscr{R}$ and $\mathscr{C}$ denotes a set of real and complex numbers, respectively. F denotes an arbitrary field (although in this book only relevant fields are $\mathscr{R}$ and $\mathscr{C}$, all statements containing F are valid for an arbitrary field). $F^{m\times n}$

denotes set of $m \times n$ matrices over the field F. Identity matrix of the format $n \times n$ will be denoted by I_n where diagonal matrix whose diagonal entries are $d_1, d_2, \ldots, d_n$ will be denoted by $diag(d_1, d_2, \ldots, d_n)$. Zero matrix of any format will be denoted by O. With $*$, we will denote conjugate transpose operation.

We will show some basic definitions and theorems from linear algebra and theory of matrices which will be used throughout the book. First, we will define few basic classes of matrices.

Definition 2.1.1. *The square matrix $A \in \mathscr{C}^{n\times n}$ ($A \in \mathscr{R}^{n\times n}$) is*

- *Hermitian (symmetric) if $A^* = A$ ($A^T = A$),*
- *normal if $A^*A = AA^*$ ($A^TA = AA^T$),*
- *unitary (orthogonal) if $A^* = A^{-1}$ ($A^T = A^{-1}$),*
- *lower triangular if $a_{ij} = 0$ for $i > j$,*
- *upper triangular if $a_{ij} = 0$ for $i < j$, or*
- *positive semi-definite if* Re $(x^*Ax) \geq 0$ *for every $x \in \mathscr{C}^{n\times 1}$. Additionally, if holds* Re $(x^*Ax) > 0$ *for every $x \in \mathscr{C}^{n\times 1} \setminus \{\mathbb{O}\}$, matrix A is called positive definite.*

Next lemma shows one important property of Hermitian positive definite matrices.

Lemma 2.1.1. *Let $A \in \mathscr{C}^{n\times n}$ be a Hermitian and positive definite matrix. Then, there exists matrix $B \in \mathscr{C}^{n\times n}$, such that $A = B^*B$. Matrix B is called square root of A and denoted by $B = A^{1/2}$.*

Definition 2.1.2. *For any matrix $A \in \mathscr{C}^{m\times n}$, define null space $N(A)$ as the inverse image of $\mathbb{O}$ vector, i.e.,*

$$N(A) = \{x \in \mathscr{C}^{n\times 1} \mid Au = \mathbb{O}\}.$$

Also, define the range $R(A)$ as the set of all images

$$R(A) = \{y \in \mathscr{C}^{m\times 1} \mid y = Ax \text{ for some } x \in \mathscr{C}^{n\times 1}\}.$$

The dimension of range $R(A)$ is called *rank* of the matrix A and denoted by rank$A$. Also, $F_r^{m\times n}$, denotes all $m \times n$ matrices A over F whose rank is rank$A = r$.

Another important characteristic of matrices is index.

Proposition 2.1.2. *For every $A \in \mathscr{C}^{n\times n}$ there exists an integer k such that* $\text{rank}A^{k+1} = \text{rank}A^k$.

Definition 2.1.3. *Let $A \in \mathscr{C}^{n\times n}$. Smallest integer k such that holds* $\text{rank}(A^{k+1}) = \text{rank}(A^k)$ *is called the index of A and is denoted by* $\text{ind}(A) = k$.

Note that if A is regular, then $\text{ind}(A) = 0$ and otherwise $\text{ind}(A) \geq 1$. Matrix index plays an important role in studying Drazin inverse.

Next, famous matrix decompositions will be used in the future considerations.

Lemma 2.1.3. *For any matrix $A \in \mathscr{C}^{m\times n}$, there exist unitary matrices $Q \in \mathscr{C}^{m\times m}$ and $P \in \mathscr{C}^{n\times n}$ such that holds $A = Q^*RP$, where matrix R has the form*

$$R = \begin{bmatrix} R_{11} & \mathbb{O} \\ \mathbb{O} & \mathbb{O} \end{bmatrix} \in \mathscr{C}^{m\times n}, \quad R_{11} \in \mathscr{C}_k^{k\times k}. \tag{2.1.1}$$

The special case of the decomposition in Lemma 2.1.3 is the famous singular value decomposition (SVD) given in the following theorem.

Theorem 2.1.4. (Singular value decomposition) *Let $A \in \mathscr{C}^{m\times n}$. There exists unitary matrices $U \in \mathscr{C}^{m\times m}$ and $V \in \mathscr{C}^{n\times n}$ and matrix*

$$\Sigma = \begin{bmatrix} diag(\sigma_1, \sigma_2, \ldots, \sigma_k) & \mathbb{O} \\ \mathbb{O} & \mathbb{O} \end{bmatrix} \in \mathscr{C}^{m\times n},$$

such that holds $A = U^*\Sigma V$. *Such decomposition is called SVD of matrix A.*

Another notable decompositions are LU and Cholesky factorization, given in the following theorem.

Theorem 2.1.5. (LU and Cholesky factorization) *For every regular square matrix* $A \in \mathscr{C}^{n\times n}$, *there exists lower triangular matrix L and upper triangular matrix U such that* $A = LU$ *and* $l_{ii} = 1$ *for every* $i = 1,2,\ldots,n$. *This factorization is known as LU factorization. Moreover, if A is Hermitian and positive definite, holds* $U = L^*$ *and factorization* $A = LL^*$ *is called Cholesky factorization.*

We will also mention the full-rank factorization.

Definition 2.1.4. (Full-rank factorization) *Let* $A \in \mathscr{C}^{m\times n}$ *be an arbitrary matrix. Factorization* $A = FG$, *where F is full-column rank matrix and G is full-row rank matrix is called full-rank factorization of matrix A.*

Theorem 2.1.6. *Let* $A \in \mathscr{C}^{m\times n}$ *such that A is of neither full-column rank nor full-row rank. Then, there exists at least one full-rank factorization* $A = FG$ *of matrix A.*

At the end of this subsection, we will state the well-known Jordan normal form.

Theorem 2.1.7. (Jordan normal form) *Let* $A \in \mathscr{C}^{n\times n}$. *There exists regular matrix P such that* $A = PJP^{-1}$, *where* $J = diag(J_1,\ldots,J_k)$ *and each block* J_i *has the form*

$$J_i = \begin{bmatrix} \lambda & 1 & 0 & \cdots & 0 & 0 \\ 0 & \lambda & 1 & \cdots & 0 & 0 \\ \cdots & \cdots & \cdots & \cdots & \cdots & \cdots \\ 0 & 0 & 0 & \cdots & \lambda & 1 \\ 0 & 0 & 0 & \cdots & 0 & \lambda \end{bmatrix},$$

and λ *is the eigenvalue of A.*

2.1.1 Computing $\{i, j, \ldots, k\}$ inverses of constant matrices

The following theorem is proved by Penrose in 1955, and it is often used as the definition of the MP inverse.

Theorem 2.1.8. *For any matrix $A \in \mathscr{R}^{m\times n}$, the following system of matrix equations*

$$\begin{array}{ll} (1)\; AXA = A, & (3)\; (AX)^T = AX, \\ (2)\; XAX = X, & (4)\; (XA)^T = XA \end{array} \tag{2.1.2}$$

has unique solution $X \in \mathscr{R}^{n\times m}$. This solution is known as MP inverse (generalized inverse) of matrix A and denoted by $A^\dagger$.

If A is a square regular matrix, then its inverse matrix A^{-1} trivially satisfies system (2.1.2). It follows that the MP inverse of a nonsingular matrix is the same as the ordinary inverse, i.e., $A^\dagger = A^{-1}$. Equations (2.1.2) are called *Penrose equations* and are used for the definition of other classes of generalized inverses.

Definition 2.1.5. *Let $A\{i, j, \ldots, k\}$ be set of matrices satisfying equations $(i), (j), \ldots, (k)$ among $(1), \ldots, (4)$ from (2.1.2). Such matrices are called $\{i, j, \ldots, k\}$ inverses and denoted by $A^{(i,j,\ldots,k)}$. The set of all $A^{(i,j,\ldots,k)}$ is denoted by $A\{i, j, \ldots, k\}$.*

One of the main properties of MP inverse is characterization of minimum-norm least-squares solution of linear system $Ax = b$. This characterization is also proven by Penrose in 1955.

Definition 2.1.6. *Least-squares solution of the linear system $Ax = b$ where $A \in \mathscr{C}^{m\times n}$ and $b \in \mathscr{C}^{m\times 1}$ is vector $u \in \mathscr{C}^{n\times 1}$ such that for any $v \in \mathscr{C}^{n\times 1}$ holds $\|Au - b\| \leq \|Av - b\|$. The least-squares solution u is minimum-norm least-squares solution if for any other least-square solution u' holds $\|u\| \leq \|u'\|$.*

Theorem 2.1.9. *Let $A \in \mathscr{C}^{m\times n}$ and $b \in \mathscr{C}^{m\times 1}$. Minimum-norm least-squares solution of the system $Ax = b$ is given by $x^* = A^\dagger b$. All other least-squares solutions are given by*

$$x = A^\dagger b + (I_n - A^\dagger A)z, \quad z \in \mathscr{C}^{n\times 1}.$$

For the consistent linear systems $Ax = b$, its minimum-norm solutions are characterized by the following theorem.

Theorem 2.1.10. [3, 80] *For $A \in \mathscr{C}^{m\times n}$ holds: for every $b \in R(A)$, $AXb = b$ and $\|Xb\| < \|u\|$ for any other solution $u \neq Xb$ of system $Ax = b$, if and only if $X \in A\{1,4\}$.*

On the other hand, all least-squares solutions of the inconsistent linear system $Ax = b$ are characterized by the following theorem.

Theorem 2.1.11. [3, 80] *For $A \in \mathscr{C}^{m\times n}$ holds: Xb is least-squares solution of $Ax = b$ for every $b \notin R(A)$ if and only if $X \in A\{1,3\}$.*

The following lemma lists some basic properties of MP and $\{1\}$ inverses.

Lemma 2.1.12. *Let $A \in \mathscr{C}^{m\times n}$. Then*

(1) $(A^\dagger)^\dagger = A$, $(A^\dagger)^* = (A^*)^\dagger$;

(2) $(\lambda A)^\dagger = \lambda^\dagger A^\dagger$, *where* $\lambda \in \mathscr{C}$ *and* $\lambda^\dagger = \begin{cases} \frac{1}{\lambda}, & \lambda \neq 0 \\ 0, & \lambda = 0 \end{cases}$;

(3) $(AA^*)^\dagger = (A^*)^\dagger A^\dagger$, $(A^*A)^\dagger = A^\dagger (A^*)^\dagger$;

(4) $A^\dagger AA^* = A^* = A^*AA^\dagger$;

(5) $A^\dagger = (A^*A)^\dagger A^* = A^*(AA^*)^\dagger$;

(6) $N(AA^\dagger) = N(A^\dagger) = N(A^*) = R(A)$;

(7) $R(AA^*) = R(AA^{(1)}) = R(A)$, $\text{rank}(AA^{(1)}) = \text{rank}(A^{(1)}A) = \text{rank}A$;

(8) *If* $\text{rank}A = m$, *then* A^*A *is regular and* $A^\dagger = (A^*A)^{-1}A^*$.

Some additional properties of $A^\dagger$ and $A^{(1)}$ not mentioned in Lemma 2.1.12 can be found, for example, in Refs. [3, 80]. In the following three lemmas and theorems, we will give the representation of $A^\dagger$ using the decomposition given in Lemma 2.1.3, SVD (Theorem 2.1.4), and full-rank factorization.

Lemma 2.1.13. *Let* $A \in \mathscr{C}^{m\times n}$ *is an arbitrary matrix. Consider the decomposition* $A = Q^*RP$ *from Lemma 2.1.3. Then, MP inverses of matrices* R *and* A, $R^\dagger$ *and* $A^\dagger$ *can be represented by*

$$R^\dagger = \begin{bmatrix} R_{11}^{-1} & \mathbb{O} \\ \mathbb{O} & \mathbb{O} \end{bmatrix}, \quad A^\dagger = Q^*R^\dagger P. \tag{2.1.3}$$

As the special case of the previous lemma is representation of $A^\dagger$ by the singular value decomposition of the matrix A.

Theorem 2.1.14. *Let* $A \in \mathscr{C}^{m\times n}$ *and let* $A = U\Sigma V^*$ *be the SVD of* A. *If*

$$\Sigma = \begin{bmatrix} diag(\sigma_1, \sigma_2, \ldots, \sigma_k) & \mathbb{O} \\ \mathbb{O} & \mathbb{O} \end{bmatrix} \in \mathscr{C}^{m\times n},$$

then

$$\Sigma^\dagger = \begin{bmatrix} diag(1/\sigma_1, 1/\sigma_2, \ldots, 1/\sigma_k) & \mathbb{O} \\ \mathbb{O} & \mathbb{O} \end{bmatrix} \in \mathscr{C}^{n\times m}$$

and $A^\dagger = V\Sigma^\dagger U^*$.

Theorem 2.1.15. *Let* $A \in \mathscr{C}^{m\times n}$ *and* $A = FG$ *its full-rank factorization. Then holds*

$$A^\dagger = G^*(F^*AG^*)^{-1}F^* = G^*(GG^*)^{-1}(F^*F)^{-1}F^*.$$

The following theorem is one intermediate result which will be used in the Subsection 4.1 in consideration of the feedback compensation problem. As far as author knows, this result is original and therefore it is given with the complete proof.

Theorem 2.1.16. *Let $A,B \in \mathscr{C}^{m\times n}$ be an arbitrary matrix. Then $A = AB^{(1)}B$ and $B = AA^{(1)}B$ holds if and only if $R(A) = R(B)$ and $R(A^*) = R(B^*)$.*

Proof.

($\Leftarrow$:) Using Lemma 2.1.12, we have that rank$A$ which is equal to rank$(AB^{(1)}B) \leq$ rank$(B^{(1)}B) =$ rank$B$ and similarly holds (using $B = AA^{(1)}B$) that rank$B \leq$ rank$A$. This proves that rank$A =$ rank$B$. On the other side, $R(B) = R(AA^{(1)}B) \subseteq R(AA^{(1)}) = R(A)$ (Lemma 2.1.12) and hence rank$B =$ rank$A$, and there holds $R(A) = R(B)$. Analogously using $A^* = (B^{(1)}B)^*A^*$ and $R((B^{(1)}B)^*) = R(B^*)$ (Lemma 2.1.12), we prove $R(A^*) = R(B^*)$.

($\Rightarrow$:) Let $x \in \mathscr{C}^{n\times 1}$ be an arbitrary vector. Since $R(A) = R(B)$, there must exist vector $y \in \mathscr{C}^{n\times 1}$ such that $Bx = Ay$. Now, it holds $AA^{(1)}Bx = AA^{(1)}Ay = Ay = Bx$. This proves $B = AA^{(1)}B$. Relation $A = AB^{(1)}B$ can be analogously proven. $\square$

Now, we will consider one application of $\{1\}$ inverses in characterization of the solution of some matrix equation.

Theorem 2.1.17. *Let $A \in \mathscr{C}^{m\times n}, B \in C^{p\times q}$, and $D \in \mathscr{C}^{m\times q}$. Then, the matrix equation*

$$AXB = D \tag{2.1.4}$$

is consistent if and only if, for an arbitrary $A^{(1)}$ and $B^{(1)}$ holds

$$AA^{(1)}DB^{(1)}B = D, \tag{2.1.5}$$

in which case the general solution is

$$X = A^{(1)}DB^{(1)} + Y - A^{(1)}AYBB^{(1)}. \tag{2.1.6}$$

Next corollary yields directly from Theorem 2.1.17 by putting $B = I_p$.

Corollary 2.1.18. *Let $A \in \mathscr{C}^{m\times n}, D \in \mathscr{C}^{m\times p}$. Matrix equation $AX = D$ has the solution if and only if $AA^{(1)}D = D$ for every $\{1\}$ inverse $A^{(1)}$.*

The following theorem considers a system of two matrix equations.

Theorem 2.1.19. *Let $A \in \mathscr{C}^{p\times m}$, $B \in \mathscr{C}^{p\times n}$, $D \in \mathscr{C}^{n\times q}$, and $E \in \mathscr{C}^{m\times q}$. The matrix equations*

$$AX = B, \quad XD = E$$

have a common solution if and only if each equation separately has ta solution and holds $AE = BD$. In such case, general solution is

$$X = X_0 + (I_m - A^{(1)}A)Y(I_n - DD^{(1)})$$

where $A^{(1)}$ and $D^{(1)}$ are an arbitrary $\{1\}$ inverses, $Y \in \mathscr{C}^{m\times n}$ is an arbitrary matrix, and

$$X_0 = A^{(1)}B + ED^{(1)} - A^{(1)}AED^{(1)}.$$

2.1.2 Computing Weighted Moore–Penrose Inverse

In the previous subsection, the relations between the generalized inverses $A^{(1,4)}, A^{(1,3)}$, and $A^{\dagger}$, and the minimum-norm solution, least-squares solution, and minimum-norm least-squares solution are discussed (Theorems 2.1.9, 2.1.10, and 2.1.11). In these cases, minimization was considered under the usual inner product $(x,y) = y^*x$ and norm $\|x\| = (x,x)^{1/2}$.

We may want to study different weighted norms for the solution x and for the residual $Ax - b$ of the linear system $Ax = b$.

Let $A \in \mathscr{C}^{m\times n}$ and let $M \in \mathscr{C}^{m\times m}$ and $N \in \mathscr{C}^{n\times n}$ be Hermitian positive definite matrices. Weighted inner products in spaces $\mathscr{C}^m$ and $\mathscr{C}^n$ can be defined by

$$(x,y)_M = y^*Mx, \quad (x,y)_N = y^*Nx.$$

According to these scalar products, weighted norms $\|x\|_M$ and $\|x\|_N$ can be defined as usual. Let us recall that conjugate transpose matrix A^* satisfies $(Ax,y) = (x,A^*y)$ for every $x \in \mathscr{C}^{m\times 1}$ and $y \in \mathscr{C}^{n\times 1}$. In the same manner, we can define *weighted conjugate transpose matrix* $A^\#$ which satisfies $(Ax,y)_M = (x,A^\#)_N$. Following lemma gives the characterization of $A^\#$.

Lemma 2.1.20. *For every $M \in \mathscr{C}^{m\times m}$ and $N \in \mathscr{C}^{n\times n}$, Hermitian positive definite matrices, and an arbitrary matrix $A \in \mathscr{C}^{m\times n}$, there exists unique matrix $A^\#$ and holds $A^\# = N^{-1}A^*M$.*

The relations between the weighted generalized inverses and the solutions of linear equations are given in the following theorem.

Theorem 2.1.21. *Let $A \in \mathscr{C}^{m\times n}$ and N be a Hermitian positive definite matrix of order n. Then, $x = Xb$ is the minimum-norm (according to norm $\| \ \|_N$) solution of the consistent system of linear equations $Ax = b$ for any $b \in R(A)$ if and only if X satisfies*

$$(1)\ AXA = A; \quad (4N)\ (NXA)^* = NXA. \tag{2.1.7}$$

Every matrix X satisfying (2.1.7) is called $\{1,4N\}$ inverse and denoted by $A^{(1,4N)}$. Set of all $A^{(1,4N)}$ is denoted by $A\{(1,4N)\}$.

Theorem 2.1.22. *Let $A \in \mathscr{C}^{m\times n}$ and M be a Hermitian positive definite matrix of order m. Then, $x = Xb$ is the least-squares (according to the norm $\| \ \|_M$) solution of the inconsistent system of linear equations $Ax = b$ for every $b \notin R(A)$ if and only if X satisfies*

$$(1)\ AXA = A; \quad (3M)(MAX)^* = MAX. \tag{2.1.8}$$

Every matrix X satisfying (2.1.8) is called $\{1,3M\}$ inverse and denoted by $A^{(1,3M)}$. Set of all $A^{(1,3M)}$ is denoted by $A\{(1,3M)\}$.

Theorem 2.1.23. *Let $A \in \mathscr{C}^{m\times n}$ and M,N be a Hermitian positive definite matrices of orders m and n, respectively. Then, $x = Xb$ is the minimum-norm (according to the norm $\|\ \|_N$) least-squares (according to the norm $\|\ \|_M$) solution of the inconsistent system of linear equations $Ax = b$ for every $b \notin R(A)$ if and only if X satisfies*

$$\begin{aligned} &(1)\ AXA = A, && (3M)\ (MAX)^* = MAX, \\ &(2)\ XAX = X, && (4N)\ (NXA)^* = NXA. \end{aligned} \tag{2.1.9}$$

Moreover, system of matrix equations (2.1.9) has a unique solution.

A matrix X satisfying (2.1.9) is called *the weighted MP inverse*, and is denoted by $X = A^{\dagger}_{MN}$. The weighted MP inverse $A^{\dagger}_{MN}$ is the generalization of MP inverse $A^{\dagger}$. If $M = I_m$, and $N = I_n$, then $A^{\dagger}_{MN} = A^{\dagger}$. Some properties of $A^{\dagger}_{MN}$ are given as follows.

Lemma 2.1.24. *Let $A \in \mathscr{C}^{m\times n}$. If $M \in \mathscr{C}^{m\times m}$ and $N \in \mathscr{C}^{n\times n}$ are Hermitian positive definite matrices, then holds*

(1) $(A^{\dagger}_{MN})^{\dagger}_{NM} = A$,

(2) $(A^{\dagger}_{MN})^* = (A^*)^{\dagger}_{N^{-1}M^{-1}}$,

(3) $A^{\dagger}MN = (A^*MA)^{\dagger}_{I_mN}A^*M = N^{-1}A^*(AN^{-1}A^*)^{\dagger}_{MI_n}$,

(4) *If $A = FG$ is a full-rank factorization of A, then* $A^{\dagger}_{MN} = N^{-1}G^*(F^*MAN^{-1}G^*)^{-1}F^*M$,

(5) $A^{\dagger}_{MN} = N^{-1/2}(M^{-1/2}AN^{-1/2})^{\dagger}M^{-1/2}$.

2.1.3 Computing the Drazin Inverse

In the previous two subsections, we discussed the MP inverse and the other $\{i, j, \ldots, k\}$ inverses which possess some 'inverse like' properties. The $\{i, j, \ldots, k\}$ inverses provide some type of solution, or least-square solution, for a system of linear equations just as inverse provides a unique solution for a nonsingular system of linear equations. Hence, the $\{i, j, \ldots, k\}$ inverses are called equation-solving inverses.

However, we also show that there are some properties of inverse matrix that the $\{i, j, \ldots, k\}$ inverses do not possess. For example, $A^{-}A = AA^{-}$, $(A^{-})^{p} = (A^{p})^{-}$, etc. The Drazin inverse and its special case group inverse will possess all of the mentioned properties. These generalized inverses are defined only for square matrix, but there are some extensions on the rectangular matrices (see for example [3]) .

Theorem 2.1.25. *Let $A \in \mathscr{C}^{n\times n}$ be an arbitrary matrix and let $k = \operatorname{ind}(A)$. Then, the following solution of matrix equations*

$$(1^k)\ A^kXA = A^k, \quad (2)\ XAX = X, \quad (5)\ AX = XA, \tag{2.1.10}$$

has a unique solution. This solution is called Drazin inverse of matrix A and denoted by A^D.

The following lemma gives the basic properties of Drazin inverse.

Lemma 2.1.26. *Let $A \in \mathscr{C}^{n\times n}$ and let $k = \operatorname{ind}(A)$. Then holds*

(1) $(A^*)^D = (A^D)^*$, $(A^T)^D = (A^D)^T$, $(A^n)^D = (A^D)^n$ *for any* $n = 1, 2, \ldots$,

(2) $((A^D)^D)^D = A^D$, $(A^D)^D = A$ *if and only if* $k = 1$,

(3) $R(A^D) = R(A^l)$ *and* $N(A^D = N(A^l)$ *for every* $l \geq k$,

(4) *If* λ *is an eigenvalue of A then,* $\lambda^{\dagger}$ *is an eigenvalue of* A^D.

Drazin inverse can be computed from the Jordan normal form of the matrix A. This is given by the following theorem.

Theorem 2.1.27. *Let $A \in \mathscr{C}^{n\times n}$ have the Jordan normal form*

$$A = PJP^{-1} = P\begin{bmatrix} J_1 & \mathbb{O} \\ \mathbb{O} & J_0 \end{bmatrix} P^{-1}, \tag{2.1.11}$$

where J_1 and J_0 contain blocks with nonzero and zero eigenvalues. Then holds

$$A^D = P^{-1}\begin{bmatrix} J_1^{-1} & \mathbb{O} \\ \mathbb{O} & \mathbb{O} \end{bmatrix} P. \tag{2.1.12}$$

Note that more general result holds. If A is decomposed in the form

$$A = PJP^{-1} = P\begin{bmatrix} C & \mathbb{O} \\ \mathbb{O} & N \end{bmatrix} P^{-1},$$

where C is regular and N is nilpotent matrix (there exists n such that $A^n = \mathbb{O}$), then

$$A^D = P^{-1}\begin{bmatrix} C^{-1} & \mathbb{O} \\ \mathbb{O} & \mathbb{O} \end{bmatrix} P.$$

Decomposition (2.1.11) is the special case of (2.1.12).

Additional properties of Drazin inverse and definitions and properties of other spectral inverses are given, for example, in Refs. [3, 80].

2.2 Methods for Computing Generalized Inverses of Constant Matrices

In this section, we will restate some known methods for computing various generalized inverses of matrices. Some of these methods (or ideas which they are based on) are already introduced in the previous section. Note that there are quite a lot of different methods for computing various classes of generalized inverses. For the brief summary of these methods, see for example [3, 80].

2.2.1 Methods Based On Full-Rank Factorization

We will state the several representations of generalized inverses, expressed in terms of full-rank factorizations and adequately selected matrices.

Let us recall that the representation of MP inverse in terms of full-rank factorization is given in Theorem 2.1.15 in the previous section. Similar result for the weighted MP inverse is given in Lemma 2.1.24.

The following lemma introduces two general representations of $\{1,2\}$ inverses, given in Refs. [48].

Theorem 2.2.1. *Let $A = PQ$ be a full-rank factorization of $A \in \mathscr{C}_r^{m\times n}$. Let U, V be matrices that run through the set of the matrices of the type $m \times m$ and $n \times n$, respectively.*

$$\begin{aligned} S_1 = \{&VQ^*(P^*UAVQ^*)^{-1}P^*U \mid \quad U \in \mathbb{C}^{m\times m},\ V \in \mathbb{C}^{n\times n},\\ &rank(P^*UAVQ^*) = r\} \\ S_2 = \{&W_1(W_2AW_1)^{-1}W_2 \mid \quad W_1 \in \mathbb{C}^{n\times r},\ W_2 \in \mathbb{C}^{r\times n},\\ &rank(W_2AW_1) = r\} \end{aligned}$$

satisfy $S_1 = S_2 = A\{1,2\}$.

Note that in reference book of Radic, a general solution of the system of matrix equations (1), (2) of (2.1.2) in the following form is introduced:

$$X = W_1(QW_1)^{-1}(W_2P)^{-1}W_2 = W_1(W_2AW_1)^{-1}W_2.$$

Here, $A = PQ$ is an arbitrary full-rank factorization and W_1, W_2 are run through the set of the matrices of the type $n \times r$ and $r \times m$, respectively, which satisfy the conditions $rank(QW_1) = rank(W_2P) = rank(A)$.

Also, we point out the following general representations for $\{1,2,3\}$ and $\{1,2,4\}$ inverses.

Theorem 2.2.2. *If $A = PQ$ is a full-rank factorization of A and W_1, W_2 satisfy the conditions from Theorem 2.2.1, then:*

(1) *General solution of the system of matrix equations (1), (2), (3) is*

$$W_1(QW_1)^{-1}(P^*P)^{-1}P^* = W_1(P^*AW_1)^{-1}P^* \ , \qquad (2.2.13)$$

(2) *General solution of the system of equations (1), (2), (4) is*

$$Q^*(QQ^*)^{-1}(W_2P)^{-1}W_2 = Q^*(W_2AQ^*)^{-1}W_2 \ . \qquad (2.2.14)$$

The general representation for the class of $\{2\}$-inverses, given in Ref. [62] is introduced by the following lemma.

Theorem 2.2.3. *The set of $\{2\}$-inverses of a given matrix $A \in \mathbb{C}_r^{m\times n}$ is determined by the following equality:*

$$A\{2\} = \{W_1(W_2AW_1)^{-1}W_2,\ W_1 \in \mathbb{C}^{n\times t},\ W_2 \in \mathbb{C}^{t\times m},\ rank(W_2AW_1)$$
$$= t,\ t = 1,\ldots,r\}.$$

At the end, we will state the following full-rank representation of the Drazin inverse. The full-rank factorization is performed on the rank invariant powers A^l, $l \geq \text{ind}(A)$. This result is also given in the Ref. [62].

Theorem 2.2.4. *Let $A \in \mathbb{C}_r^{n\times n}$, $k = \text{ind}(A)$, and $l \geq k$ is an arbitrary integer. If $A^l = PQ$ is the full-rank factorization of the matrix A^l, then the Drazin inverse of A has the following representation:*

$$A^D = P(QAP)^{-1}Q. \qquad (2.2.15)$$

There are a lot of methods based on the special full-rank factorization, for example LU factorization. Presented lemmas holds also for such decompositions.

2.2.2 Leverrier–Faddeev Method

In this section, we will introduce the well-known Leverrier–Faddeev method. This algorithm has been rediscovered and modified several times.

In 1840, the Frenchman U.J.J. Leverrier provided the basic connection with Newton–Girard relations. J.M. Souriau and J.S. Frame, independently modified the algorithm to its present form. That is why Leverrier–Faddeev method is also called Souriau–Frame method. Paul Horst along with Faddeev and Sominskii are also credited with rediscovering the technique. Gower developed the algorithm further and pointed out its value, not for computation, but for deriving algebraic results. Although the algorithm is intriguingly beautiful, it is not practical for floating-point computations.

It is originally used for computing characteristic polynomial of the given matrix A. Later, on, it is shown how this method can be extended for the computing various classes of generalized inverses (MP, Drazin, and various other generalized inverses). For the sake of completeness, all these well-known results are restated, and the proof of some of them is also given.

2.2.3 Method of Zhukovski

In Zhukovski method starting from the known recurrent equations for solving linear system $Ax = y$, where $A \in \mathbb{C}_m^{m\times n}$,

$$x_{t+1} = x_t + \frac{\gamma_t a_{t+1}^*}{a_{t+1}\gamma_t a_{t+1}^*}\left(y_{t+1} - a_{t+1}x_t\right), \quad x_0 = \mathbb{O},$$

$$\gamma_{t+1} = \gamma_t - \frac{\gamma_t a_{t+1}^* a_{t+1}\gamma_t}{a_{t+1}\gamma_t a_{t+1}^*}, \quad \gamma_0 = I_n.$$

In the same work, he introduced the generalization of these recurrent relations, and thus found a solution for consensual system of linear algebraic equations $Ax = y$, in the case of the matrix $A \in \mathbb{C}_r^{m\times n}$, $r \le m \le n$. It is shown that the X_T, γ_t, $t = 1,\ldots,m$ is the solution of next system of equations

$$x_{t+1} = x_t + \gamma_t a_{t+1}^* \left(a_{t+1}\gamma_t a_{t+1}^*\right)^\dagger (y_{t+1} - a_{t+1}x_t), \quad x_0 = \mathbb{O},$$
$$\gamma_{t+1} = \gamma_t - \gamma_t a_{t+1}^* \left(a_{t+1}\gamma_t a_{t+1}^*\right)^\dagger a_{t+1}\gamma_t, \quad \gamma_0 = I_n,$$

where

$$\left(a_{t+1}\gamma_t a_{t+1}^*\right)^\dagger = \begin{cases} \frac{1}{a_{t+1}\gamma_t a_{t+1}^*}, & a_{t+1}\gamma_t a_{t+1}^* > 0, \\ 0, & a_{t+1}\gamma_t a_{t+1}^* = 0. \end{cases}$$

Note that the $a_{t+1}\gamma_t a_{t+1}^* = 0$ if and only if type a_{t+1} is linearly dependent on the type of $a_1,\ldots,a_t$.

From here, he developed a method to calculate the inverse matrix MP $A \in \mathbb{C}^{m\times n}$, $m \leq n$.

Let Γ_t, $t = 1,\ldots,n$ be an array $n \times n$ of matrix, defined with

$$\Gamma_{t+1} = \Gamma_t - \Gamma_t b_{t+1} \left(b_{t+1}^*\Gamma_t b_{t+1}\right)^\dagger b_{t+1}^*\gamma_t, \quad \gamma_0 = I_n. \tag{2.2.16}$$

Now with $b_t, t = 1,\ldots,n$ denotes the column matrix A, while C_t, $t = 1,\ldots,n$ are the kinds of matrix $I_n - \Gamma_n$. Consider a string matrix X_t and γ_t, defined as follows

$$\begin{aligned} \gamma_{t+1} &= \gamma_t - \gamma_t a_{t+1}^* \left(a_{t+1}\gamma_t a_{t+1}^*\right)^\dagger a_{t+1}\gamma_t, \quad \gamma_0 = I_n, \\ X_{t+1} &= X_t + \gamma_t a_{t+1}^* \left(a_{t+1}\gamma_t a_{t+1}^*\right)^\dagger (c_{t+1} - a_{t+1}x_t), \quad X_0 = \mathbb{O}. \end{aligned} \tag{2.2.17}$$

Next theorem connects a set of pre-defined matrices X_T, the MP inverse $A^\dagger$ matrix A, and shows the correctness constructed method for calculating the inverse MP.

Theorem 2.2.5. *Let array X_t, $t = 0,\ldots,n$ be defined as in* (2.2.16) *and* (2.2.17). *Then, $X_n = A^\dagger$.*

2.2.4 Relation Between Characteristic Polynomial and Inverse Matrix

Let us remember that for every square matrix $A \in \mathscr{C}^{n\times n}$, its *characteristic polynomial* $P_A(x)$ is defined by

$$P_A(x) = \det[xI_n - A].$$

Now, we will recall two very important theorems which provides Leverrier–Faddeev method.

Theorem 2.2.6. (Newton–Girard relations) *Let $P(x)$ be any polynomial of degree n, and let*

$$P(x) = a_0x^n + a_1x^{n-1} + \ldots + a_{n-1}x + a_n. \tag{2.2.18}$$

Denote by $x_1, x_2, \ldots, x_n$ all n roots of $P(x)$ and $\sigma_k = \sum_{i=1}^{n} x_i^k$ for every $k = 1,2,\ldots,n$. Then, the following relation

$$ka_k + \sigma_1 a_{k-1} + \ldots + \sigma_{k-1}a_1 + \sigma_k a_0 = 0, \tag{2.2.19}$$

is satisfied for every $k = 1,2,\ldots,n$.

Now, let $A \in \mathscr{C}^{n\times n}$ be an arbitrary square matrix. Consider the characteristic polynomial $P_A(x)$ in the form (2.2.18). From the Theorem 2.2.6 (relation (2.2.19)), we can write coefficient a_k in the form

$$a_k = -\frac{1}{k}\left(\sigma_k + \sigma_{k-1}a_1 + \ldots + \sigma_1 a_{k-1}\right). \tag{2.2.20}$$

Now, $\sigma_k = \sum_{i=1}^{n} \lambda_i^k = \mathrm{tr}(A^k)$ for $k = 1,2,\ldots,n$, where $\lambda_1, \lambda_2, \ldots, \lambda_n$ are eigenvalues of matrix A. By replacing in Eq. (2.2.20), we obtain

$$a_k = -\frac{1}{k}\,\mathrm{tr}(A^k + a_1A^{k-1} + \ldots + a_{k-1}A) \tag{2.2.21}$$

Denote for every $k = 1,2,\ldots,n$,

$$B_k = A^k + a_1A^{k-1} + \ldots + a_{k-1}A + a_kI_n.$$

and

$$A_k = AB_{k-1} \tag{2.2.22}$$

Then, relation (2.2.21) can be written as

$$a_k = -\frac{1}{k}Tr(A_k). \tag{2.2.23}$$

The following relation trivially holds

$$B_k = B_{k-1} + a_k I_n. \tag{2.2.24}$$

All coefficients a_k, $k = 1, 2, \ldots, n$ of the characteristic polynomial $P_A(x)$ can be computed by successive application of the relations (2.2.22), (2.2.23), and (2.2.24). Hence, we can establish Algorithm 2.1, and by the previous consideration, the following theorem holds.

Theorem 2.2.7. *For any matrix* $A \in \mathscr{C}^{m \times n}$, *the result of Algorithm 2.1 is the characteristic polynomial* $P_A(x)$ *of matrix* A.

Algorithm 2.1 Leverrier–Faddeev method for computing the characteristic polynomial $P_A(x)$ of a given matrix A

Input: an arbitrary square matrix $A \in \mathscr{C}^{n \times n}$.

$a_0 := 1$
$A_0 := \mathbb{O}$
$B_0 := I_n$
for $i := 1$ to n **do**
    $A_i := AB_{i-1}$
    $a_i := -\operatorname{tr}(A_i)/i$
    $B_i := A_i + a_i I_n$
end for
return $P_A(x) := a_0 x^n + a_1 x^{n-1} + \ldots + a_{n-1}x + a_n$

2.2.5 Computing the Moore–Penrose Inverse of Constant Matrices

Now, we will derive an algorithm based on Algorithm 2.1 for computing MP inverse of the given matrix A. The following famous result will be useful.

Theorem 2.2.8. (Cayley–Hamilton) *For every matrix $A \in \mathscr{C}^{n\times n}$ holds $P_A(A) = \mathbb{O}$, i.e., matrix A is root of its characteristic polynomial.*

Decell showed that MP inverse $A^{\dagger}$ can be directly obtained from characteristic polynomial $P_{AA^*}(x)$. Following theorem gives us the connection between characteristic polynomial $P_{AA^*}(x)$ of matrix AA^* and generalized inverse $A^{\dagger}$. For the sake of completeness, we will also give the proof of Decell's result. This proof is different than the one discussed in Decel's result.

Theorem 2.2.9. *Let $A \in \mathscr{C}^{m\times n}$ and*

$$P_{AA^*}(x) = \det[xI_n - AA^*] = a_0x^n + a_1x^{n-1} + \ldots + a_{n-1}x + a_n, \quad a_0 = 1,$$

Let k be a maximal index such that $a_k \neq 0$ (i.e., if $a_k \neq 0$ and $a_{k+1} = \ldots = a_n = 0$). If $k > 0$, then MP inverse of matrix A is given by

$$A^{\dagger} = -a_k^{-1}A^*[(AA^*)^{k-1} + a_1(AA^*)^{k-2} + \ldots + a_{k-2}AA^* + a_{k-1}] \quad (2.2.25)$$

Else if $k = 0$, then holds $A^{\dagger} = \mathbb{O}$.

Proof. If $k = 0$, then $A = \mathbb{O}$ and hence $A^{\dagger} = \mathbb{O}$. Let us suppose that $k \neq 0$.

Direct application of the Cayley–Hamilton theorem to the matrix AA^* yields

$$(AA^*)^n + a_1(AA^*)^{n-1} + \ldots + a_{k-1}(AA^*)^{n-k+1} + a_k(AA^*)^{n-k} = \mathbb{O}. \quad (2.2.26)$$

Consider the decomposition of A (according to Lemma 2.1.3, such decomposition exists) in the form $A = Q^*R'P$, where $Q \in \mathscr{C}^{n\times n}$ and $P \in \mathscr{C}^{n\times n}$ are unitary matrices and R' is given by

$$R' = \begin{bmatrix} R'_{11} & \mathbb{O} \\ \mathbb{O} & \mathbb{O} \end{bmatrix}, \quad R'_{11} \in \mathscr{C}_k^{k\times k}. \tag{2.2.27}$$

Then holds $AA^* = Q^*RQ$ where $R = R'R'^*$ can be decomposed on the same way as R in (2.2.27)

$$R = \begin{bmatrix} R_{11} & \mathbb{O} \\ \mathbb{O} & \mathbb{O} \end{bmatrix}, \quad R_{11} = R'_{11}R'^*_{11} \in \mathscr{C}_k^{k\times k}. \tag{2.2.28}$$

Since $(AA^*)^i = Q^*R^iQ$ for every $i = 1, 2, 3, \ldots$, by replacing into (2.2.26), we obtain

$$Q^*\left[R^n + a_1R^{n-1} + \ldots + a_{k-1}R^{n-k+1} + a_kR^{n-k}\right]Q = \mathbb{O}.$$

Since Q is regular and R has the form (2.2.28), it holds

$$R_{11}^n + a_1R_{11}^{n-1} + \ldots + a_{k-1}R_{11}^{n-k+1} + a_kR_{11}^{n-k} = \mathbb{O}.$$

Multiplying by $R_{11}^{-(n-k)}$ yields to

$$R_{11}^k + a_1R_{11}^{k-1} + \ldots + a_{k-1}R_{11} + a_kI_k = \mathbb{O}. \tag{2.2.29}$$

Denote

$$I_{m,k} = \begin{bmatrix} I_k & \mathbb{O} \\ \mathbb{O} & \mathbb{O} \end{bmatrix} \in \mathscr{C}^{m\times m}.$$

Equation (2.2.29) remains true when R_{11} is replaced with R and I_k by $I_{m,k}$.

$$R^k + a_1R^{k-1} + \ldots + a_{k-1}R + a_kI_{m,k} = \mathbb{O}. \tag{2.2.30}$$

Multiplying (2.2.30) by Q^* from left and Q from right side, respectively, we obtain

$$(AA^*)^k + a_1(AA^*)^{k-1} + \ldots + a_{k-1}(AA^*) + a_kQ^*I_{m,k}Q = \mathbb{O}. \tag{2.2.31}$$

Last equation is equivalent with

$$-a_k^{-1}AA^*[(AA^*)^{k-1}+a_1(AA^*)^{k-2}+\ldots+a_{k-2}(AA^*)+a_{k-1}]=Q^*I_mQ. \tag{2.2.32}$$

Since $A^\dagger$ can be expressed in terms of Q,P, and $R^\dagger$ by Lemma 2.1.13

$$A^\dagger = P^*R^\dagger Q, \quad R^\dagger = \begin{bmatrix} R_{11}^{-1} & \mathbb{O} \\ \mathbb{O} & \mathbb{O} \end{bmatrix}, \tag{2.2.33}$$

by multiplying both sides of (2.2.32) by $A^\dagger$ and using $A^\dagger AA^* = A^*$ (Lemma 2.1.12) we obtain statement of the theorem. □

Now from the Theorem 2.2.9 and Algorithm 2.1, an algorithm for computing MP inverse can be established (Algorithm 2.2).

Algorithm 2.2 Leverrier–Faddeev method for computing MP inverse $A^\dagger$ of a given matrix A

Input: Matrix $A \in \mathscr{C}^{m\times n}$.

```
a_0 := 1
A_0 := O
B_0 := I_n
for i := 1 to n do
    A_i := AA* B_{i-1}
    a_i := - tr(A_i)/i
    B_i := A_i + a_i I_n
end for
k := max{i | a_i ≠ 0, i = 0,...,n}
if k = 0 then
    return A† := O
else
    return A† := a_k^{-1} A* B_{k-1}
end if
```

Let us perform the complexity analysis of Algorithm 2.2. The body of the loop given in steps 4–8 is repeated n times. It consists of two matrix–matrix multiplications, one trace computation, and matrix–matrix addition. The complexity of the whole body is $\mathcal{O}(n^3)$ (assuming that matrix–matrix multiplication runs in $\mathcal{O}(n^3)$). Therefore, complexity of loop in steps 4–8 is $\mathcal{O}(n^3 \cdot n) = \mathcal{O}(n^4)$.

Step 9 requires $\mathcal{O}(n)$ time, and step 13 requires $\mathcal{O}(n^3)$ time. Therefore, the total complexity of Algorithm 2.2 is $\mathcal{O}(n^4)$.

First method in this class is Greville's partitioning method [21] for computing the MP inverse. The main idea is to express $A_k^\dagger$ as the function of $A_{k-1}^\dagger$, A_{k-1}, and a_k. For $k = 2, \ldots, n$, let the vectors d_k and c_k be defined by

$$\begin{aligned} d_k &= A_{k-1}^\dagger a_k \\ c_k &= a_k - A_{k-1} d_k = a_k - A_{k-1} A_{k-1}^\dagger a_k \end{aligned} \tag{2.2.34}$$

Required connection between $A_k^\dagger$ on one and $A_{k-1}^\dagger$, A_{k-1}, and a_k on the other side is given by the following theorem.

Theorem 2.2.10. [3, 21] *Let $A \in \mathcal{C}^{m \times n}$. Using the above notation, the MP inverse of A_k $(k = 2, 3, \ldots, n)$ is*

$$A_k^\dagger = \begin{bmatrix} A_{k-1} & a_k \end{bmatrix}^\dagger = \begin{bmatrix} A_{k-1}^\dagger - d_k b_k^* \\ b_k^* \end{bmatrix}, \tag{2.2.35}$$

where

$$b_k^* = \begin{cases} c_k^\dagger, & c_k \neq \mathbb{O} \\ (1 + d_k^* d_k)^{-1} d_k^* A_{k-1}^\dagger, & c_k = \mathbb{O} \end{cases} \tag{2.2.36}$$

Based on Theorem 2.2.10, Algorithm 2.3 for computation of MP inverse can be constructed. Initial calculation of $A_1^\dagger = a_1^\dagger$ can be done as follows

$$a_1^\dagger = \begin{cases} (a_1^* a_1)^{-1} a_1^*, & a_1 \neq \mathbb{O} \\ \mathbb{O}, & a_1 = \mathbb{O} \end{cases}. \tag{2.2.37}$$

Algorithm 2.3 Partitioning method for computing MP inverse of a given matrix A

Input: an arbitrary matrix $A \in \mathscr{C}^{m\times n}$.

$$A_1^{\dagger} := a_1^{\dagger} := \begin{cases} (a_1^* a_1)^{-1} a_1^*, & a_1 \neq \mathbb{O} \\ \mathbb{O}, & a_1 = \mathbb{O} \end{cases}$$

for $k := 2$ to n **do**

$d_k := A_{k-1}^{\dagger} a_k$

$c_k := a_k - A_{k-1} d_k$

if $c_k = \mathbb{O}$ **then**

$b_k^* := (c_k^* c_k)^{-1} c_k^*$

else

$b_k^* := (1 + d_k^* d_k)^{-1} d_k^* A_{k-1}^{\dagger}$

end if

$$A_k^{\dagger} := \begin{bmatrix} A_{k-1}^{\dagger} - d_k b_k^* \\ b_k^* \end{bmatrix},$$

end for

return $A^{\dagger} := A_n^{\dagger}$

As it was done for the previous algorithms, we will perform the complexity analysis of Algorithm 2.3. Consider the loop in the steps 2–11. The most time consuming operation is matrix–vector multiplication in steps 3, 4, and 8 and computation of the product $d_k b_k^*$ in step 10. Complexity of these operations is $\mathscr{O}(m \cdot k)$. Therefore, total complexity of the whole Algorithm 2.3 is

$$\mathscr{O}\left(\sum_{k=2}^{n} m \cdot k\right) = \mathscr{O}(m \cdot n^2).$$

There are also other variants of Algorithm 2.3. For example, Cline's method [9, 80] computes MP inverse of partitioned matrix $A = [U \ V]$.

Also, Noble's method generalizes both Greville and Cline method and under some suitable conditions compute MP and Drazin inverse of block matrix,

$$A = \begin{bmatrix} A_{11} & A_{12} \\ A_{21} & A_{22} \end{bmatrix}. \tag{2.2.38}$$

2.2.6 Drazin Inverse

Similarly as in the case of MP inverse, the Drazin inverse A^D can be computed using the characteristic polynomial $P_A(x)$ of matrix A. In Grevile and Ji reference books, the following representation of the Drazin inverse is introduced.

Theorem 2.2.11. *Consider a square matrix $A \in \mathscr{C}^{n\times n}$. Assume that*

$$P_A(x) = \det[xI_n - A] = a_0x^n + a_1x^{n-1} + \cdots + a_{n-1}x + a_n, \quad a_0 = 1,$$

Consider the following sequence of $n \times n$ constant matrices defined by coefficients a_i and powers of A:

$$B_j = a_0A^j + a_1A^{j-1} + \cdots + a_{j-1}A + a_jI_n, \quad a_0 = 1, \quad j = 0,\ldots,n \tag{2.2.39}$$

Let r denote the smallest integer such that $B_r = \mathbb{O}$, let t denote the largest integer satisfying $a_t \neq 0$, and let $k = r - t$. Then, the Drazin inverse A^D of A is given by

$$A^D = (-1)^{k+1} a_t^{-k-1} A^k B_{t-1}^{k+1}. \tag{2.2.40}$$

Now from Theorem 2.2.11, the following extension of Leverrier–Faddeev method for computing Drazin inverse yields a straightforward algorithm (Algorithm 2.4).

Let us perform the complexity analysis of Algorithm 2.4. Similarly to the case of Algorithm 2.2, loop in steps 4–8 has the complexity $\mathscr{O}(n^4)$. Step 10 requires $\mathscr{O}(n \cdot n^2)$ time. In step 12, we have to compute r-th and $r+1$-th

Algorithm 2.4 Leverrier–Faddeev method for computing Drazin inverse A^D of a given square matrix A

Input: Matrix $A \in \mathscr{C}^{n \times n}$.

$a_0 := 1$
$A_0 := \mathbb{O}$
$B_0 := I_n$
for $i := 1$ to n **do**
  $A_i := AB_{i-1}$
  $a_i := -\operatorname{tr}(A_i)/i$
  $B_i := A_i + a_i I_n$
end for
$k := \max\{i \mid a_i \neq 0, i = 0, \ldots, n\}$
$t := \min\{i \mid B_i = \mathbb{O}, i = 0, \ldots, n\}$
$r := t - k$
return $A^D := (-1)^{r+1} a_k^{-r-1} A^r B_{k-1}^{r+1}$

matrix power of matrices A and B_{k-1}, respectively. This matrix powering can be done in $\mathscr{O}(n^3 \log n)$ because $r \leq n$.

Therefore, the complexity of Algorithm 2.4 is equal to the $\mathscr{O}(n^4)$, and it is the same as the complexity of Algorithm 2.4 for computing MP inverse.

2.2.7 Other Generalized Inverses

In the previous two subsection are shown two extensions of Leverrier–Faddeev method which computes MP and Drazin inverse of an input matrix, respectively. This idea can be further generalized such that we can establish a general extension of Leverrier–Faddeev method (Leverrier–Faddeev type algorithm) for computing various classes of generalized inverses.

This is done by Algorithm 2.5. In contract to the Algorithm 2.2 and Algorithm 2.4, which take only one input matrix A whose inverse is being calculated, Algorithm 2.5 takes two matrices R and T and one natural number $e \in \mathcal{N}$. Method is restated from Ref. Pecko.

Algorithm 2.5 General Leverrier–Faddeev type algorithm for computing various generalized inverses

Input: Matrices $R, T \in \mathscr{C}^{n\times m}$ and positive integer $e \in \mathcal{N}$.

$a_0 := 1$
$A_0 := \mathbb{O}$
$B_0 := I_n$
for $i := 1$ to n **do**
  $A_i := TR^* B_{i-1}$
  $a_i := -\operatorname{tr}(A_i)/i$
  $B_i := A_i + a_i I_n$
end for
$k := \max\{i \mid a_i \neq 0, i = 0, \ldots, n\}$
if $k = 0$ **then**
  return $X_e := \mathbb{O}$
else
  return $X_e := (-1)^e a_k^{-e} R^* B_{k-1}^e$
end if

Next theorem (restated from Ref. Pecko) shows how to use Algorithm 2.5 for computing different types of generalized inverses of a given constant, rational, or polynomial matrix A.

Theorem 2.2.12. *Let A be an $n \times m$ constant, rational, or polynomial matrix and $A = PQ$ its full-rank factorization. The following statements are valid:*

(1) *If $R = T = A$, there holds $X_1 = A^\dagger$.*

(2) *If* $m = n$, $R^* = A^l$, $T = A$, *and* $l \geq \operatorname{ind}(A)$, *there holds* $X_1 = A^D$.

(3) *If* $T = A$ *and* $n > m = \operatorname{rank} A$ *for an arbitrary* R *such that* AR^* *is invertible, there holds* $X_1 = A_R^{-1}$.

(4) *If* $m = n$, $R^* = A^k$, *and* $T = I_n$, *then* X_1 *exists if and only if* $\operatorname{ind}(A) = k$ *and* $X_1 = AA^D$.

(5) *In the case* $m = n$, $e = l + 1$, $TR^* = A$, $R^* = A^l$, *and* $l \geq \operatorname{ind}(A)$, *there holds* $X_e = A^D$.

(6) *For* $m = n$, $e = 1$, $T = R^* = A^l$, *and* $l \geq \operatorname{ind}(A)$, *we have* $X_1 = (A^D)^l$.

(7) $X_1 \in A\{2\}$ *if and only if* $T = A, R = GH, G \in \mathscr{C}^{n\times t}, H \in \mathscr{C}^{t\times m}$, *and* $\operatorname{rank} HAG = t$.

(8) $X_1 \in A\{1,2\}$ *if and only if* $T = A, R = GH, G \in \mathscr{C}^{n\times r}, H \in \mathscr{C}^{r\times m}$, *and* $\operatorname{rank} HAG = r = \operatorname{rank} A$.

(9) $X_1 \in A\{1,2,3\}$ *if and only if* $T = A, R = GP^*, G \in \mathscr{C}^{n\times r}$, *and* $\operatorname{rank} P^*AG = r = \operatorname{rank} A$.

(10) $X_1 \in A\{1,2,4\}$ *if and only if* $T = A, R = Q^*H, H \in \mathscr{C}^{r\times n}$, *and* $\operatorname{rank} HAQ^* = r = \operatorname{rank} A$.

(11) *If* $T = A$ *and* $m > n = \operatorname{rank} A$ *for an arbitrary* R *such that* R^*A *is invertible, there holds* $X_1 = A_L^{-1}$.

By performing complexity analysis similarly as for the Algorithm 2.2 and Algorithm 2.4, it can be obtained that total complexity of Algorithm 2.5 when input matrices are constant is $\mathscr{O}(n^4 + n^3 \log e)$. This assumes that matrix–matrix multiplications are performed in $\mathscr{O}(n^3)$ time. Let us only mention that the complexity of step 13 is $\mathscr{O}(n^3 \log e)$. Throughout the rest of this section, we will additionally suppose $e \ll 10^n$, such that total complexity of Algorithm 2.5 for input matrix $A \in \mathscr{C}^{n\times m}$ is $\mathscr{O}(n^4)$.

2.3 Partitioning Method

This section deals with another class of well-known methods for computing generalized inverses. All these methods are finite iterative methods. At the k-th iteration $k = 1, 2, \dots, n$, they compute generalized inverse A_k^-, where A_k is the submatrix of A consisting of its first k columns. Hence, matrix A_k is partitioned in the following way

$$A_k = [A_{k-1} \; a_k] \tag{2.3.41}$$

where a_k is k-th column of the matrix A. That is why these methods are called *partitioning methods.*

2.3.1 $\{1\}$ Inverse

Algorithm 2.3 can also be used for the computation of $\{1\}$ inverse, since $A^\dagger \in A\{1\}$. However, if we need just $\{1\}$ inverse of A, Algorithm 2.3 can be significantly simplified. The following theorem establishes the partitioning method for computing $\{1\}$ inverse of the given matrix A.

Theorem 2.3.1. [3] *Let $A \in \mathscr{C}^{m\times n}$ be an arbitrary matrix. Then for every $k = 2, \dots, n$ holds*

$$A_k^{(1)} = \begin{bmatrix} A_{k-1}^{(1)} - d_k b_k^* \\ b_k^* \end{bmatrix},$$

where d_k and b_k^* are defined by

$$d_k = A_{k-1}^{(1)} a_k$$

$$c_k = a_k - A_{k-1} d_k$$

$$b_k^* = \begin{cases} \mathbb{O}, & c_k = \mathbb{O} \\ c_k^\dagger (I_m - A_{k-1} A_{k-1}^\dagger), & c_k \neq \mathbb{O} \end{cases}$$

Time complexity of algorithm based on Theorem 2.3.1 can be similarly derived as in the case of Algorithm 2.3 and it is also equal to $\mathscr{O}(n^3)$.

2.3.2 Weighted Moore–Penrose Inverse

Similarly to the Greville's method, it can be constructed the method for computing the weighted MP inverse of partitioned matrices. This is done by Miao, Wang and Chen [81].

We will consider positive definite Hermitian matrices $M \in \mathscr{C}^{m\times m}$ and $N \in \mathscr{C}^{n\times n}$. The leading principal submatrix $N_i \in \mathscr{C}^{i\times i}$ of N is partitioned as

$$N_i = \begin{bmatrix} N_{i-1} & l_i \\ l_i^* & n_{ii} \end{bmatrix}, \; i = 2,\ldots,n, \tag{2.3.42}$$

where $l_i \in \mathscr{C}^{(i-1)\times 1}$ and n_{ii} is the complex polynomial. By N_1 we denote the polynomial n_{11}.

For the sake of simplicity, by X_i we denote the weighted MP inverse $X_i = (A_i)^{\dagger}_{MN_i}$, for each $i = 2,\ldots,n$. Similarly, $X_1 = (a_1^{\dagger})_{M,N_1}$.

Theorem 2.3.2. [81] *Let $A \in \mathscr{C}^{m\times n}$, and assume that $M \in \mathscr{C}^{m\times m}$, and $N \in \mathscr{C}^{n\times n}$ are positive definite Hermitian matrices. Then for every $i = 2,3,\ldots,n$, matrix X_i can be computed in the following way*

$$X_1 = \begin{cases} (a_1^* M a_1)^{-1} a_1^* M, & a_1 \neq 0, \\ a_1^*, & a_1 = 0, \end{cases} \tag{2.3.43}$$

$$X_i = \begin{bmatrix} X_{i-1} - (d_i + (I_n - X_{i-1}A_{i-1}) N_{i-1}^{-1} l_i) b_i^* \\ b_i^* \end{bmatrix} \tag{2.3.44}$$

where the vectors d_i, c_i, and b_i^* are defined by

$$d_i = X_{i-1} a_i \tag{2.3.45}$$

$$c_i = a_i - A_{i-1} d_i = (I - A_{i-1} X_{i-1}) a_i \tag{2.3.46}$$

$$b_i^* = \begin{cases} (c_i^* M c_i)^{-1} c_i^* M, & c_i \neq 0 \\ \delta_i^{-1} (d_i^* N_{i-1} - l_i^*) X_{i-1}, & c_i = 0, \end{cases} \tag{2.3.47}$$

and δ_i is

$$\delta_i = n_{ii} + d_i^* N_{i-1} d_i - (d_i^* l_i + l_i^* d_i) - l_i^* (I - X_{i-1} A_{i-1}) N_{i-1}^{-1} l_i. \tag{2.3.48}$$

Also in Ref. [81], authors used a block representation of the inverse N_i^{-1}, which we also generalized to the set of rational matrices.

Lemma 2.3.3. [**81**] *Let N_i be the partitioned matrix defined in* (2.3.42). *Assume that N_i and N_{i-1} are both nonsingular. Then for every $k = 2, 3, \ldots, n$ holds*

$$N_i^{-1} = \begin{cases} \begin{bmatrix} N_{i-1} & l_i \\ l_i^* & n_{ii} \end{bmatrix}^{-1} = \begin{bmatrix} E_{i-1} & f_i \\ f_i^* & h_{ii} \end{bmatrix}, & i = 2, \ldots, n, \\ n_{11}^{-1}, & i = 1, \end{cases} \tag{2.3.49}$$

where

$$h_{ii} = \left(n_{ii} - l_i^* N_{i-1}^{-1} l_i\right)^{-1}, \tag{2.3.50}$$

$$f_i = -h_{ii} N_{i-1}^{-1} l_i, \tag{2.3.51}$$

$$E_{i-1} = N_{i-1}^{-1} + h_{ii}^{-1} f_i f_i^*. \tag{2.3.52}$$

In view of Theorem 3.2.1 and Lemma 3.2.2, respectively, we present the following algorithms for computing the weighted MP inverse and the inverse matrix $N_i^{-1} \in \mathscr{C}^{i \times i}$.

2.4 Matrix Multiplication is as Hard as Generalized Inversion

This section deals with the new method for computing MP and some other $\{i, j, \ldots, k\}$ inverses, based on the generalized Cholesky decomposition [11]. A recursive algorithm for generalized Cholesky factorization of a given symmetric, positive semi-definite matrix $A \in \mathscr{C}^{n \times n}$ is introduced. This algorithm works in matrix multiplication time. Strassen methods for matrix multiplication and inversion are used together with the recursive generalized Cholesky factorization algorithm, establishing an algorithm for

---

Algorithm 2.1 Partitioning method for computing MP inverse of a given multivariable rational or polynomial matrix.

Input: Matrix $A \in \mathscr{C}^{m\times n}$ and positive definite matrices $M \in \mathscr{C}^{m\times m}$ and $N \in \mathscr{C}^{n\times n}$.

```
Compute X_1 = a_1^† defined in (2.3.43)
for i = 2 to n do
    Compute d_i using (2.3.45)
    Compute c_i using (2.3.46)
    Compute b_i^* by means of (2.3.47) and (2.3.48)
    Applying (2.3.44) compute X_i
end for
return X_n
```

Algorithm 2.2 Computing inverse of a given symmetric positive-definite multivariable rational or polynomial matrix.

Input: Symmetric, positive definite matrix $N \in \mathscr{C}^{n\times n}$

```
N_1^{-1} := n_{11}^{-1}
for i := 2 to n do
    Compute h_ii using (2.3.50)
    Compute f_i using (2.3.51)
    Compute E_{i-1} using (2.3.52)
    Compute N_i^{-1} using (2.3.49)
end for
return N_n^{-1}
```

computing MP, $\{1,2,3\}$, $\{1,2,4\}$, $\{2,3\}$, and $\{2,4\}$ inverses. Introduced algorithms are not harder than the matrix–matrix multiplication.

This section presents the original results and is based on the paper Dexter Chol.

2.4.1 Strassen's Inversion Method

Let $A,B \in \mathscr{C}^{n\times n}$. The usual number of scalar operations required for computing matrix product $C = AB$ using the well-known formulas

$$c_{ij} = \sum_{k=1}^{n} a_{ik}b_{kj},$$

is $2n^3 - n^2 = \mathscr{O}(n^3)$ (n^3 multiplications and $n^3 - n^2$ additions). In the V. Strassen's paper, he introduced the method for matrix multiplication, the complexity of which is $\mathscr{O}(n^{\log_2 7}) \approx \mathscr{O}(n^{2.807})$. Strassen method is recursive and based on the following proposition.

Proposition 2.4.1. [10] *Let the matrices $A,B \in \mathscr{C}^{2n\times 2n}$ and $C = AB$ be partitioned as*

$$A = \begin{bmatrix} A_{11} & A_{12} \\ A_{21} & A_{22} \end{bmatrix}, \quad B = \begin{bmatrix} B_{11} & B_{12} \\ B_{21} & B_{22} \end{bmatrix}, \quad C = \begin{bmatrix} C_{11} & C_{12} \\ C_{21} & C_{22} \end{bmatrix},$$

and all blocks have dimensions $n \times n$. Denote

1. $Q_1 = (A_{11} + A_{22})(B_{11} + B_{22})$
2. $Q_2 = (A_{21} + A_{22})B_{11}$
3. $Q_3 = A_{11}(B_{12} - B_{22})$
4. $Q_4 = A_{22}(B_{21} - B_{11})$
5. $Q_5 = (A_{11} + A_{12})B_{22}$
6. $Q_6 = (A_{21} - A_{11})(B_{11} + B_{12})$
7. $Q_7 = (A_{12} - A_{22})(B_{21} + B_{22})$

(2.4.53)

Then, we can express blocks of matrix C in the following way

$$\begin{aligned} C_{11} &= Q_1 + Q_4 - Q_5 + Q_7 & C_{12} &= Q_3 + Q_5 \\ C_{21} &= Q_2 + Q_4 & C_{22} &= Q_1 + Q_3 - Q_2 + Q_6 \end{aligned} \tag{2.4.54}$$

Using the Proposition, 2.4.1 we have to multiply seven matrices of the format $n \times n$ in order to obtain product of two $2n \times 2n$ matrices. These matrix products are computed by the recursion. Algorithm 2.1 realizes this procedure.

Let us denote the complexity of Algorithm 2.3 by $\text{inv}(n)$. Also, denote by $\text{add}(n)$ the complexity of the matrix addition on $n \times n$ matrices and by $\text{mul}(m,n,k)$ the complexity of multiplying $m \times n$ matrix with $n \times k$ matrix.

Time complexity of Algorithm 2.1 can be determined using the well-known Master Theorem. Formulation and proof of the Master Theorem can be found, for example, in Ref. [10]. Since we have to apply the same method seven times on the matrices with half dimension and other operations are matrix additions, it holds

$$\text{mul}(n,n,n) = 7 \cdot \text{mul}(n/2,n/2,n/2) + \mathcal{O}(n^2). \tag{2.4.55}$$

By Master Theorem, solution of recurrence (2.4.55) is $\text{mul}(n,n,n) = \mathcal{O}(n^{\log_2 7})$. In general case holds

$$\text{mul}(m,n,k) = \mathcal{O}((\max\{m,n,k\})^{\log_2 7}).$$

Algorithm 2.1 Strassen multiplication

Input: Matrices $A, B \in \mathcal{C}^{n \times n}$.

if $n = 1$ **then**

  return $C = [c_{11}] := [a_{11} \cdot b_{11}]$

end if

if n is odd **then**

  Adopt matrices A and B by adding one zero row and column.

  $n_{old} := n$

  $n := n + 1$

end if

Algorithm 2.2 Strassen multiplication (Continued...)

Make the block decomposition $\begin{bmatrix} A_{11} & A_{12} \\ A_{21} & A_{22} \end{bmatrix}, B = \begin{bmatrix} B_{11} & B_{12} \\ B_{21} & B_{22} \end{bmatrix}$ where blocks have dimensions $n/2 \times n/2$.

Compute values $Q_1, \ldots, Q_7$ using (2.4.53) where matrix products are computed recursively.

Compute values C_{11}, C_{12}, C_{21}, and C_{22} using (2.4.54).

$C := \begin{bmatrix} C_{11} & C_{12} \\ C_{21} & C_{22} \end{bmatrix}$

if n_{old} is odd **then**

Drop the last (zero) row and column of matrix C

end if

return C

There are other algorithms for computing the product $C = AB$ in time below $\mathscr{O}(n^3)$. Currently, the best one is due to Coppersmith and Winograd and it works in time $\mathscr{O}(n^{2.376})$.

Strassen also introduced the algorithm for the inversion of given $n \times n$ matrix A with the same complexity as the matrix multiplication. This algorithm is also based on the block decomposition of the matrix A and following lemma.

Lemma 2.4.2. *If A is a given $n \times n$ matrix partitioned in the following way*

$$A = \begin{bmatrix} A_{11} & A_{12} \\ A_{21} & A_{22} \end{bmatrix}, \quad A_{11} \in \mathscr{C}^{k\times k} \tag{2.4.56}$$

and both A and A_{11} are regular, then the inverse matrix $X = A^{-1}$ can be represented in the well-known form of the block matrix inversion

$$X = \begin{bmatrix} X_{11} & X_{12} \\ X_{21} & X_{22} \end{bmatrix} = \begin{bmatrix} A_{11}^{-1} + A_{11}^{-1}A_{12}S^{-1}A_{21}A_{11}^{-1} & -A_{11}^{-1}A_{12}S^{-1} \\ -S^{-1}A_{21}A_{11}^{-1} & S^{-1} \end{bmatrix}. \tag{2.4.57}$$

Matrices X_{11}, X_{12}, X_{21} *and* X_{22} *in the block form* (2.4.57) *can be computed with minimal number of matrix multiplications if we use intermediate matrices* $R_1, \ldots, R_7$ *defined by the following relations*

$$
\begin{array}{llll}
1. & R_1 = A_{11}^{-1} & 7. & X_{12} = R_3 R_6 \\
2. & R_2 = A_{21} R_1 & 8. & X_{21} = R_6 R_2 \\
3. & R_3 = R_1 A_{12} & 9. & R_7 = R_3 X_{21} \\
4. & R_4 = A_{21} R_3 & 10. & X_{11} = R_1 - R_7 \\
5. & R_5 = R_4 - A_{22} & 11. & X_{22} = -R_6. \\
6. & R_6 = R_5^{-1} & &
\end{array}
\qquad (2.4.58)
$$

Let us notice that the matrix R_5 in the relations (2.4.58) is equal to the minus Schur complement of A_{11} in the matrix A

$$R_5 = -(A_{22} - A_{21} A_{11}^{-1} A_{12}) = -S = -(A/A_{11}).$$

Formulas (2.4.57) and (2.4.58) are applicable if both A_{11} and the Schur complement $S = (A/A_{11})$ are invertible. Formulas (2.4.58) can be used for recursive computation of matrix inverse A^{-1}. Relations (1) and (6) in (2.4.58) contain matrix inverses of matrices with lower dimension (i.e., $k \times k$ and $(n-k) \times (n-k)$, respectively). By applying the same formulas recursively on these submatrices, it is obtained the recursive method for matrix inversion. In such case, recursion is continued down to the case of 1×1 matrices. Algorithm 2.3 is the realization of such method.

It is known that Algorithm 2.3 for the inversion has the same complexity as matrix multiplication, i.e., $\mathrm{inv}(n) = \Theta(\mathrm{mul}(n)) = \Theta(n^{2+\varepsilon})$, under the assumption $\mathrm{add}(n) = \mathcal{O}(n^2)$ and $\mathrm{mul}(n) = \Theta(n^{2+\varepsilon})$ where $0 < \varepsilon < 1$, [10].

Remark 2.4.1. If any algorithm for matrix–matrix multiplication with complexity $\mathcal{O}(n^{2+\varepsilon})$ is used, then Algorithm 2.3 also works with complexity $\mathcal{O}(n^{2+\varepsilon})$, $0 < \varepsilon < 1$ (again by the Master Theorem). Especially,

Algorithm 2.3 Strassen-based matrix inversion

Input: Matrix $A \in \mathscr{C}^{n \times n}$ such that all diagonal minors $A_{(S)}$ are regular.

1: **if** $n = 1$ **then**
2: **return** $X := [a_{11}^{-1}]$
3: **end if**
4: $k := \lfloor n/2 \rfloor$
5: Make the block decomposition $A = \begin{bmatrix} A_{11} & A_{12} \\ A_{21} & A_{22} \end{bmatrix}$ such that $A_{11} \in \mathscr{C}^{k \times k}$.
6: Compute values $R_1, \ldots, R_6$ and $X_{11}, X_{12}, X_{21}, X_{22}$ using (2.4.58) where inverses are computed recursively.
7: **return** $X = \begin{bmatrix} X_{11} & X_{12} \\ X_{21} & X_{22} \end{bmatrix}$

if the Strassen's matrix–matrix multiplication algorithm (Algorithm 2.1) is applied, Algorithm 2.3 requires

$$\frac{6}{5} n^{\log_2 7} - \frac{1}{5} n \approx n^{2.807}$$

multiplications. Otherwise, if the usual matrix–matrix multiplication algorithm with ordinary time complexity $\mathscr{O}(n^3)$ is used, then complexity of Algorithm 2.3 is $\mathscr{O}(n^3)$.

CHAPTER 3

GENERALIZED INVERSES OF POLYNOMIAL AND RATIONAL MATRICES

In this chapter, we will consider algorithms for computing generalized inverses of polynomial and rational matrices. At the end of the chapter, we have given some applications of generalized inverses of constant, rational, and polynomial matrices.

This section reviews some important properties of polynomial and rational matrices which will be used later. Let F be the an arbitrary field. We can restrict this notation just to the fields $\mathscr{R}$ or $\mathscr{C}$, but all considerations in this section holds also in the case of an arbitrary field F.

The $m \times n$ matrix $A = [a_{ij}]$ is *polynomial* (*rational*) if all its elements a_{ij} are polynomials (rational functions) of the single variable s. Such matrices will be denoted by $A(s)$. Similarly multivariable polynomial (rational) matrices are defined and denoted by $A(s_1, \ldots, s_p)$ (if the variables are $s_1, \ldots, s_p$). In this case, we will also use the shorter notation $A(S)$ where $S = (s_1, \ldots, s_p)$. The set of all polynomial (rational) $m \times n$ matrices $A(s)$ will be denoted by $F^{m\times n}[s]$ ($F^{m\times n}(s)$). Similar notation will be used for multivariable rational ($F^{m\times n}(s_1,\dots,s_p) = F^{m\times n}(S)$) and polynomial ($F^{m\times n}[s_1,\dots,s_p] = F^{m\times n}[S]$) matrices.

Following definitions consider the degree of polynomial matrix $A(s)$.

Definition 3.0.1. *For a given polynomial matrix $A(s) \in F[s]^{n\times m}$, its maximal degree is defined as the maximal degree of its elements*

$$\deg A(s) = \max\{\mathrm{dg}(A(s))_{ij} \,|\, 1 \le i \le n, 1 \le j \le m\}. \tag{3.0.1}$$

Similarly for the given multivariable polynomial matrix $A(s_1,\dots,s_p)$, maximal degree $\deg_k A(S)$ is given by

$$\deg_k A(S) = \max\{\mathrm{dg}_k(A(S))_{ij} \,|\, 1 \le i \le n, 1 \le j \le m\}, \tag{3.0.2}$$

where $\mathrm{dg}_k P(s)$ is maximal degree of variable s_k in polynomial $P(S)$.

Definition 3.0.2. *The degree matrix corresponding to $A(s) \in F[s]^{n\times m}$ is the matrix defined by $\mathrm{dg}A(s) = [\mathrm{dg}A(s)_{ij}]_{m\times n}$.*

Next lemma shows some basic properties of degree matrices.

Lemma 3.0.1. *Let $A(s), B(s) \in F^{n\times n}[s]$ and $a(s) \in F[s]$. The following facts are valid*

(a) $\mathrm{dg}(A(s)B(s))_{ij} = \max\{\mathrm{dg}A(s)_{ik} + \mathrm{dg}B(s)_{kj} \,|\, 1 \le k \le n\}$.

(b) $\mathrm{dg}(A(s)+B(s))_{ij} \le \max\{\mathrm{dg}A(s)_{ij}, \mathrm{dg}B(s)_{ij}\}$.

(c) $\mathrm{dg}(a(s)A(s))_{ij} = \mathrm{dg}A(s)_{ij} + \mathrm{dg}(a(s))$.

Proof.

(a) From the definition of the matrix product, and using simple formulae

$$\mathrm{dg}(p(s)+q(s)) \le \max\{\mathrm{dg}(p(s)), \mathrm{dg}(q(s))\},$$

$$\mathrm{dg}(p(s)q(s)) = \mathrm{dg}(p(s)) + \mathrm{dg}(q(s))$$

for every $p(s), q(s) \in F(s)$ we conclude:

$$\mathrm{dg}(A(s)B(s))_{ij} = \mathrm{dg}((A(s)B(s))_{ij}) \leq$$

$$\max\{\mathrm{dg}A(s)_{ik} + \mathrm{dg}B(s)_{kj} \,|\, k = 1, \ldots, n\}.$$

This completes the proof of part (a). The other two parts can be similarly verified. □

Every polynomial matrix $A(s) \in F^{m\times n}[s]$ can be written as

$$A(s) = A_0 + A_1 s + \ldots + A_d s^d,$$

where $d = \deg A(s)$ and $A_i \in F^{m\times n}$ are *coefficient matrices*. Similarly, the multivariable polynomial matrix $A(s_1, \ldots, s_p) \in F^{m\times n}\ [s_1, \ldots, s_p]$ can be written in the form

$$A(S) = \sum_{i_1=0}^{d_1} \sum_{i_2=0}^{d_2} \cdots \sum_{i_p=0}^{d_p} A_{i_1 i_2 \cdots i_p} s_1^{i_1} s_2^{i_2} \cdots s_p^{i_p} = \sum_{I=0}^{D} A_I S^I.$$

Here, we denoted $D - (d_1, \ldots, d_p)$, $I = (i_1, \ldots, i_p)$, $S^I = s_1^{i_1} s_2^{i_2} \cdots s_p^{i_p}$, $A_I = A_{i_1 i_2 \cdots i_p} \in F^{m\times n}$, and $\sum_{I=0}^{D} = \sum_{i_1=0}^{d_1} \sum_{i_2=0}^{d_2} \cdots \sum_{i_p=0}^{d_p}$.

The following two definitions deal with the sparsity of the matrix. These variables define how many non-zero elements are in the matrix and will be used for the test matrices generation.

Definition 3.0.3. *For a given matrix $A(s) \in \mathscr{R}^{m\times n}[s]$,* **the first sparse number** *$sp_1(A)$ is the ratio of the total number of non-zero elements and total number of elements in $A(s)$, i.e.,*

$$sp_1(A(s)) = \frac{\#\{(i,j) \,|\, a_{ij}(s) \neq 0\}}{m \cdot n}.$$

The first sparse number represents density of non-zero elements, and it is between 0 and 1.

Definition 3.0.4. *For a given non-constant polynomial matrix* $A(s) \in F[s]^{m\times n}$, **the second sparse number** $sp_2(A(s))$ *is the following ratio*

$$sp_2(A(s)) = \frac{\#\{(i,j,k)\,|\,\mathrm{Coef}(a_{ij}(s),s^k) \neq 0\}}{\deg A(s)\cdot m\cdot n},$$

where $\mathrm{Coef}(P(s),s^k)$ *denotes coefficient corresponding to* s^k *in polynomial* $P(s)$.

The second sparse number represents density of non-zero coefficients contained in elements $a_{ij}(s)$, and it is also between 0 and 1.

We will illustrate the computation of sparse numbers $sp_1(A(s))$ and $sp_2(A(s))$ in one example.

Example 3.0.1. *Consider the polynomial matrix*

$$A(s) = \begin{bmatrix} s & s^2 & 1+s+s^2 \\ 0 & 0 & 1 \\ 1 & s^2 & 0 \end{bmatrix}.$$

We have that

$$\begin{aligned}
&\#\{(i,j)\mid a_{ij}(s)\neq 0\} = \\
&\#\{(1,1),(1,2),(1,3),(2,3),(3,1),(3,2)\} = 6, \\
&\#\left\{(i,j,k)\mid \mathrm{Coef}(a_{ij}(s),s^k)\neq 0\right\} = \\
&\#\{(1,1,1),(1,2,2),(1,3,0),(1,3,1), \\
&\quad (1,3,2),(2,3,0),(3,1,0),(3,2,2)\} = 8.
\end{aligned}$$

Hence, sparse numbers $sp_1(A(s))$ *and* $sp_2(A(s))$ *are equal to*

$$sp_1(A(s)) = \frac{6}{3\cdot 3} = \frac{2}{3}, \qquad sp_2(A(s)) = \frac{8}{3\cdot 3\cdot 2} = \frac{4}{9}.$$

Last two definitions can be naturally generalized to the multivariable polynomial matrices.

Definition 3.0.5. *For a given matrix $A(S) = [a_{ij}(S)] \in \mathscr{C}[S]^{m\times n}$ (polynomial or constant),* **the first sparse number** $sp_1(A)$ *is the ratio of the total number of non-zero elements and total number of elements in $A(S)$*

$$sp_1(A(S)) = \frac{\#\{(i,j) \mid a_{ij}(S) \neq 0\}}{m\cdot n}.$$

Definition 3.0.6. *For a given polynomial matrix $A(S) \in \mathscr{C}[S]^{m\times n}$ and $S = (s_1,\dots,s_p)$,* **the second sparse number** $sp_2(A(S))$ *is the following ratio*

$$sp_2(A(S)) =$$

$$\frac{\#\{(i,j,k_1,\dots,k_p) \mid 0 \le k_j \le \deg_{s_j} A(S), \mathrm{Coef}(a_{ij}(S), s_1^{k_1}\cdots s_p^{k_p}) \neq 0\}}{\deg_{s_1} A(S)\cdots \deg_{s_p} A(S)\cdot m\cdot n}.$$

By $\mathrm{Coef}(P(S), s_1^{k_1}\cdots s_p^{k_p})$, *we denoted the coefficient corresponding to $s_1^{k_1}\cdots s_p^{k_p}$ in polynomial $P(S)$.*

Example 3.0.2. *Consider the following two-variable polynomial matrix $A(s_1,s_2)$*

$$A(s_1,s_2) = \begin{bmatrix} s_1+s_2 & s_1^2 & s_2^2 \\ s_1s_2 & 2s_1 & 1+s_1^2 \\ 0 & 0 & 1 \end{bmatrix}.$$

Evidently holds $\deg_1 A(s_1,s_2) = \deg_2 A(s_1,s_2) = 2$. *As in the previous example, we have that holds*

$$\begin{aligned}
&\#\{(i,j) \mid a_{ij}(s_1,s_2) \neq 0\} = \\
&\#\{(1,1),(1,2),(1,3),(2,1),(2,2),(2,3),(3,3)\} = 7, \\
&\#\left\{(i,j,k) \mid \mathrm{Coef}(a_{ij}(s), s^k) \neq 0\right\} = \\
&\#\{(1,1,1,0),(1,1,0,1),(1,2,2,0),(1,3,0,2),(2,1,1,1), \\
&(2,2,1,0),(2,3,0,0),(2,3,1,0),(2,3,2,0),(3,3,0,0)\} = 10.
\end{aligned}$$

From the previous relation we can directly compute sparse numbers

$$sp_1(A(s)) = \frac{7}{3\cdot 3} = \frac{7}{9}, \qquad sp_2(A(s)) = \frac{10}{3\cdot 3\cdot 2\cdot 2} = \frac{5}{18}.$$

3.1 Symbolic Matrix Computations

Symbolic calculation of generalized inverse is usually performed through the decomposition of the full-rank appropriate matrix. In this chapter were used of benefits of LDL^* and QDR decomposition in relation to their classical variants. The modifications were used in the sense that all the matrices are of full-rank decomposition.

Method of Ref. [69], in which LU factorization is applied for calculating the pseudo-inverse matrix, is expanded in the work [58]. An obvious goal is to make an algorithm for symbolic calculation using generalized inverse LDL^* decomposition rational matrix $A \in \mathbf{C}(x)^{m \times n}$. Motivation is derived from the benefits of using free-roots decomposition (decomposition, which does not contain roots) in the algorithms for polynomial and rational matrices. In this way, you can avoid the difficulties of calculating symbolic polynomial matrices and their inverses, arising from the emergence of elements with roots. Also, it is shown that LDL^* decomposition increases efficiency calculating the generalized inverse.

3.2 Calculation of $\{i, j, k\}$ Inverse and Generalized Inverse Matrix of Rational

Given a complex Hermitian matrix A, the following recursive relation vases matrix elements D and L, obtained from LDL^* matrix decomposition A:

$$d_{jj} = a_{jj} - \sum_{k=1}^{j-1} l_{jk} l_{jk}^* d_{kk}, \tag{3.2.1}$$

$$l_{ij} = \frac{1}{d_j}(a_{ij} - \sum_{k=1}^{j-1} l_{ik} l_{jk}^* d_{kk}), \text{ za } i > j. \tag{3.2.2}$$

The idea that the calculations carried out are enough for $j = \overline{1,r}$, where $r = \text{rank}(A)$, is given in Ref. [58]. This gives a complete factorization rank of a matrix A, where L does not contain zero-column and matrix D does not contain zero–zero-type column. Therefore, for a given matrix $A \in \mathbf{C}_r^{m\times n} = \{X \in \mathbf{C}^{m\times n} \mid rank(X) = r\}$, a decomposition of the full-rank without square root is given with $A = LDL^*$, where $L \in \mathbf{C}^{m\times r}$ and $D \in \mathbf{C}^{r\times r}$ is the corresponding diagonal matrix.

Representations $\{1,2,3\}$ and $\{1,2,4\}$-inverse introduced in the Ref. [3], modified as follows from Ref. [69]. Also, these representations are known for complex matrices, the enlarged set of rational matrix of an unknown.

Lemma 3.2.1. *Observe the matrix* $A \in \mathbf{C}(x)_r^{m\times n}$ *and let* $0 < s \leq r$, $m_1, n_1 \geq s$ *be given integers. Then, the following is valid:*

(a) $A\{2,4\}_s = \{(YA)^\dagger Y \mid Y \in \mathbf{C}(x)^{n_1\times m},\ YA \in \mathbf{C}(x)_s^{n_1\times n}\}$.

(b) $A\{2,3\}_s = \{Z(AZ)^\dagger \mid Z \in \mathbf{C}(x)^{n\times m_1},\ AZ \in \mathbf{C}(x)_s^{m\times m_1}\}$.

Analogous to representations $\{1,2,3\}$ and $\{1,2,4\}$-inverses, we derive for case $s = r = rank(A)$.

Lemma 3.2.2. *Consider the matrix* $A \in \mathbf{C}(x)_r^{m\times n}$ *and let* $m_1, n_1 \geq r$ *be fixed integers. Then, the next statements are used to represent the sets* $A\{1,2,4\}$, $A\{1,2,3\}$, *and the Moore–Penrose's inverse matrix A:*

(a) $A\{1,2,4\} = \{(YA)^\dagger Y \mid Y \in \mathbf{C}(x)^{n_1\times m},\ YA \in \mathbf{C}(x)_r^{n_1\times n}\}$.

(b) $A\{1,2,3\} = \{Z(AZ)^\dagger \mid Z \in \mathbf{C}(x)^{n\times m_1},\ AZ \in \mathbf{C}(x)_r^{m\times m_1}\}$.

(c) $A^\dagger = (A^*A)^\dagger A^* = A^*(AA^*)^\dagger$.

Now it is possible to carry out the next couple of the statements to calculate $\{1,2,3\}$ and $\{1,2,4\}$ matrix inverse of the given matrix $A \in \mathbf{C}(x)_r^{m\times n}$.

Theorem 3.2.3. [58] *Let A be rational matrix* $A \in \mathbf{C}(x)_r^{m\times n}$ *and an arbitrary* $m \times n_1$ *rational matrix R, where* $n_1 \geq r$. *Let* LDL^* *be square root free*

Cholesky factorization with full-rank of matrix $G = (R^*A)^*(R^*A)$, wherein $L \in \mathbf{C}(x)^{n\times r}$ and $D \in \mathbf{C}(x)^{r\times r}$. Then, the following the statements are valid:

$$A\{1,2,4\} = \{L(L^*GL)^{-1}L^*(R^*A)^*R^* \mid R \in \mathbf{C}(x)^{m\times n_1},$$
$$R^*A \in \mathbf{C}(x)_r^{n_1\times n}\}.$$

Proof. Consider the following expression for the calculation of Moore–Penrose's product inverse matrix A and B (i.e., $(AB)^\dagger$), which is introduced in Ref. [49]:

$$(AB)^\dagger = B^*(A^*ABB^*)^\dagger A^*. \tag{3.2.3}$$

By applying this equation for the case $A = R^*A$, $B = I$, Moore–Penros's inverse

$$(R^*A)^\dagger = ((R^*A)^*(R^*A))^\dagger (R^*A)^*. \tag{3.2.4}$$

We will use LDL^* decomposition of matrix $(R^*A)^*(R^*A) = LDL^*$. Replacing these values in Eq. (3.2.4), we get

$$(R^*A)^\dagger = (LDL^*)^\dagger (R^*A)^*. \tag{3.2.5}$$

Further, with replacing $A = L$, $B = DL^*$ in equation (3.2.3), we have

$$(LDL^*)^\dagger = (DL^*)^*(L^*LDL^*(DL^*)^*)^{-1}L^* =$$
$$LD^*(L^*LDL^*LD^*)^{-1}L^*. \tag{3.2.6}$$

Then, multiplying $(R^*A)^\dagger$ with R^* on the right side, in accordance with the last equation and the equation (1.1.3), we get

$$(R^*A)^\dagger R^* = LD^*(L^*LDL^*LD^*)^{-1}L^*(R^*A)^*R^* =$$
$$L(L^*LDL^*L)^{-1}L^*(R^*A)^*R^*.$$

Here, notice that the matrix L does not contain zero columns, L^* does not contain zero type, and D are no-zero columns and zero type. It is a matrix

L^*LDL^*L square, which implies that it is invertible. Now, the proof of the statements follows on the basis of Lemma 3.2.2, working under the **(a)**. □

Notice that if $A = LDL^*$ and B is appropriate identity matrix, then

$$(LDL^*)^\dagger = (LD^*L^*LDL^*)^{-1}(LD^*L^*)$$

This equation leads to verification of last statement theorem, and proof is completed.

Corollary 3.2.4. [58] *Let is given rational matrix $A \in \mathbf{R}(x)_r^{m\times n}$ and an arbitrary $m \times n_1$ real matrix R, where $n_1 \geq r$. Suppose that LDL^T-free root Cholesky factorization full-rank of matrix $G = (R^TA)^T(R^TA)$, wherein $L \in \mathbf{R}(x)^{n\times r}$ and $D \in \mathbf{R}(x)^{r\times r}$. Then the following statements are satisfied:*

$$A\{1,2,4\} = \left\{L(L^TGL)^{-1}L^T(R^TA)^TR^T \mid R \in \mathbf{R}^{m\times n_1}, R^TA \in \mathbf{R}(x)_r^{n_1\times n}\right\}. \tag{3.2.7}$$

Theorem 3.2.5. [58] *Let be given rational matrix $A \in \mathbf{C}(x)_r^{m\times n}$ and an arbitrary $m_1 \times n$ rational matrix T, wherein $m_1 \geq r$. Suppose that LDL^* free-root Cholesky factorization full-rank of matrix $G = (AT^*)(AT^*)^*$, where $L \in \mathbf{C}(x)^{m\times r}$ and $D \in \mathbf{C}(x)^{r\times r}$. Then the following statements are valid:*

$$A\{1,2,3\} = \left\{T^*(AT^*)^*L(L^*GL)^{-1}L^* \mid T \in \mathbf{C}(x)^{m_1\times n}, AT^* \in \mathbf{C}(x)_r^{m\times m_1}\right\}. \tag{3.2.8}$$

Proof. The proof is similar to the proof of Theorem 2.1, with the difference in the application of the statements under the **(b)** Lemma 3.2.2. Taking $A = I$ and $B = AT^*$ in equation (1.1.1), we get

$$(AT^*)^\dagger = TA^*(AT^*(AT^*)^*)^\dagger. \tag{3.2.9}$$

Taking that LDL^* full-rank decomposition of matrix $(AT^*)(AT^*)^*$, we have that

$$(AT^*)^\dagger = TA^*(LDL^*)^\dagger. \tag{3.2.10}$$

Now, the proof follows on the basis of the statements **(b)** Lemma 3.2.2, analogous to the proof of Theorem 3.2.3. Notice that in this case, m and m_1 occur n and n_1, respectively. □

Corollary 3.2.6. [58] *Let be given the rational matrix* $A \in \mathbf{R}(x)_r^{m\times n}$ *and an arbitrary* $m_1 \times n$ *real matrix* T*, where* $m_1 \geq r$*. Suppose that* LDL^T *is free root Cholesky factorization full-rank of matrix* $G = (AT^T)(AT^T)^T$*, wherein* $L \in \mathbf{R}(x)^{m\times r}$ *and* $D \in \mathbf{R}(x)^{r\times r}$*. Then the following statements are valid:*

$$A\{1,2,3\} = \{T^T(AT^T)^T L(L^T GL)^{-1} L^T \mid T \in \mathbf{R}^{m_1\times n}, AT^T \in \mathbf{R}(x)_r^{m\times m_1}\}. \tag{3.2.11}$$

Theorem 3.2.7. [58]*Let be given the following rational matrix* $A \in \mathbf{C}(x)_r^{m\times n}$*. If* LDL^* *is free root Cholesky factorization of matrix* $G = (A^*A)^*(A^*A)$*, where* $L \in \mathbf{C}(x)^{m\times r}$ *and* $D \in \mathbf{C}(x)^{r\times r}$*, then the following is valid:*

$$A^\dagger = L(L^*GL)^{-1}L^*(A^*A)^*A^*. \tag{3.2.12}$$

If LDL^* *free root Cholesky factorization of matrix* $H = (AA^*)\ (AA^*)^*$*,* $L \in \mathbf{C}(x)^{m\times r}$ *and* $D \in \mathbf{C}(x)^{r\times r}$*, then the following is satisfied:*

$$A^\dagger = A^*(AA^*)^*L(L^*HL)^{-1}L^*. \tag{3.2.13}$$

Proof. Proof follows from 3.2.2, statement (c). □

Corollary 3.2.8. [58] *Let be given rational matrix* $A \in \mathbf{R}(x)^{m\times n}$*. Suppose that* LDL^T *is free root Cholesky factorization full-rank of matrix* $G = (A^TA)^T(A^TA)$*, where* $L \in \mathbf{R}(x)^{m\times r}$ *and* $D \in \mathbf{R}(x)^{r\times r}$*. The following statement is satisfied:*

$$A^\dagger = L(L^TGL)^{-1}L^T(A^TA)^TA^T. \tag{3.2.14}$$

If LDL^T is free root Cholesky factorization of matrix $H = (AA^T)(AA^T)^T$, where $L \in \mathbf{R}(x)^{m\times r}$ and $D \in \mathbf{R}(x)^{r\times r}$, then the following is valid:

$$A^{\dagger} = A^T(AA^T)^T L(L^T HL)^{-1}L^T. \tag{3.2.15}$$

Based on the proven the statements, the following algorithm is introduced 3.1 named LDLGInverse for calculating sets $A\{1,2,4\}$ and $A\{1,2,3\}$ for given rational $m \times n$ matrix A, and for calculating Moore–Penrose's inverse of matrix A.

Algorithm 3.1 Calculating $\{i, j, k\}$ and Moore–Penrose's inverse of rational matrix—Algorithm LDLGInverse

---

Input: Rational matrix $A \in \mathbf{C}(x)_r^{m\times n}$.

if $\ddagger = \{1,2,4\}$ **then**
    Generalise matrix randomly $R \in \mathbf{C}^{m\times n_1}$, $n_1 \geq r$
    $G := (R^*A)^*(R^*A)$
else if $\ddagger = \{1,2,3\}$ **then**
    Generalise matrix randomly $T \in \mathbf{C}^{m_1\times n}$, $m_1 \geq r$
    $G := (AT^*)(AT^*)^*$
else
    $G := (A^*A)^*(A^*A)$ {this step is for MP-inverse}
end if
Calculate LDL^* factorization full-rank of matrix G.
if $\ddagger = \{1,2,4\}$ **then**
    $M = L(L^*GL)^{-1}L^*(R^*A)^*R^*$
else if $\ddagger = \{1,2,3\}$ **then**
    $M = T^*(AT^*)^*L(L^*GL)^{-1}L^*$
else
    $M = L(L^*GL)^{-1}L^*(A^*A)^*A^*$
end if
Rezultat: Resulting matrix M.

Example 3.2.1. Consider the matrix A_6 from Ref. [88], where a is some constant. For calculate $\{1,2,4\}$ and $\{1,2,3\}$ inverse of matrix A_6, constant matrix $R = T^T$ is randomly generated.

$$A_6 = \begin{bmatrix} a+5 & a+3 & a+2 & a+4 & a+3 & a+2 \\ a+3 & a+4 & a+2 & a+3 & a+3 & a+2 \\ a+2 & a+2 & a+2 & a+2 & a+2 & a+1 \\ a+4 & a+3 & a+2 & a+3 & a+3 & a+2 \\ a+3 & a+3 & a+2 & a+3 & a+2 & a+2 \\ a+2 & a+2 & a+1 & a+2 & a+2 & a \\ a+1 & a & a+1 & a+1 & a+1 & a-1 \end{bmatrix},$$

$$R = T^T = \begin{bmatrix} 1 & 5 & 3 & 1 & 2 & -1 \\ 1 & -1 & 4 & 1 & -2 & 1 \\ 3 & 1 & -3 & 1 & 2 & -1 \\ -2 & -1 & 3 & 2 & -2 & 1 \\ 1 & -1 & 3 & -1 & 2 & -1 \\ 3 & -1 & -1 & 6 & -2 & 1 \\ -3 & 4 & -3 & 2 & -2 & -1 \end{bmatrix}.$$

Results for $\{1,2,4\}$, $\{1,2,3\}$ and MP inverses of matric A_6, which are obtained applying algorithm LDLGInverse, are the following:

$$A_6^{(1,2,4)} = \begin{bmatrix} -3-a & -\frac{754+543a}{55} & \frac{479+433a}{55} & 4+a & 3+a & \frac{3(233+181a)}{55} & -\frac{589+488a}{55} \\ -3-a & -\frac{847+543a}{55} & \frac{57}{5}+\frac{433a}{55} & 3+a & 3+a & \frac{77}{5}+\frac{543a}{55} & -\frac{737+488a}{55} \\ -2-a & -2-\frac{543a}{55} & \frac{433a}{55} & 2+a & 2+a & 1+\frac{543a}{55} & -\frac{488a}{55} \\ 4+a & \frac{413+543a}{55} & -\frac{138+433a}{55} & -5-a & -3-a & -\frac{358+543a}{55} & \frac{8(31+61a)}{55} \\ 3+a & \frac{3(382+181a)}{55} & -\frac{871+433a}{55} & -3-a & -4-a & -\frac{1091+543a}{55} & \frac{981+488a}{55} \\ 2+a & \frac{695+543a}{55} & -\frac{530+433a}{55} & -2-a & -2-a & -\frac{695+543a}{55} & \frac{585+488a}{55} \end{bmatrix},$$

$$A_6^{(1,2,3)} = A_6^{\dagger} = \begin{bmatrix} -3-a & -\frac{3(3+a)}{4} & -\frac{11+5a}{4} & 4+a & 3+a & \frac{5+3a}{4} & \frac{3+a}{4} \\ -3-a & -\frac{3(2+a)}{4} & -\frac{5(2+a)}{4} & 3+a & 3+a & \frac{3(2+a)}{4} & \frac{2+a}{4} \\ -2-a & -\frac{5+3a}{4} & -\frac{3+5a}{4} & 2+a & 2+a & \frac{1+3a}{4} & \frac{3+a}{4} \\ 4+a & \frac{3(3+a)}{4} & \frac{11+5a}{4} & -5-a & -3-a & -\frac{5+3a}{4} & -\frac{3+a}{4} \\ 3+a & \frac{3(3+a)}{4} & \frac{11+5a}{4} & -3-a & -4-a & -\frac{5+3a}{4} & -\frac{3+a}{4} \\ 2+a & \frac{5+3a}{4} & \frac{7+5a}{4} & -2-a & -2-a & -\frac{5+3a}{4} & -\frac{3+a}{4} \end{bmatrix}.$$

Notice that gained $\{1,2,3\}$-inverse and Moore–Penrose's inverses of matrix A_6 are equal in the previous example.

3.2.1 Implementation Details and Results of Testing

Based on restructured iteration procedure (3.2.1), it is possible to derive the following implementation LDL^* of decomposition of an arbitrary matrix in MATHEMATICA. The following function as unique argument implies the matrix given in form of list.

```
LDLDecomposition[A_List] :=
 Module[{i, j, k, n = MatrixRank[A], m = Length[A], L, D},
  L = Table[0, {m}, {n}];
  D = Table[0, {n}, {n}];
  For[j = 1, j ≤ n, j++,
   L[[j,j]] = 1; D[[j,j]] = Simplify[A[[j,j]] - ∑_{k=1}^{j-1} (L[[j,k]])^2 D[[k,k]]];
   For[i = j + 1, i ≤ m, i++,
    L[[i,j]] = Simplify[1/D[[j,j]] (A[[i,j]] - ∑_{k=1}^{j-1} L[[i,k]] L[[j,k]] D[[k,k]])];];
  ];
  Return[{L, D}];]
```

Based on Algorithm 3.1 we have the following modulo which defines Moore–Penrose's inverse of an arbitrary matrix A.

```
LDLGInverse[A_List] :=
  Module[{L, D, G, G1, M, Y},
   G1 = Transpose[A].A;
   G = G1.G1;
   {L, D} = LDLDecomposition[G];
   M = Simplify[Inverse[Transpose[L].L.D.Transpose[L].L]];
   Y = L.M.Transpose[L].G1.Transpose[A];
   Return[Simplify[Y]]];
```

Example 3.2.2. Compare different algorithms for calculating the generalized inverse algorithm LDLGInverse. The Table 3.1 shows the execution time on different test matrices obtained by using algorithms implemented in the software package MATHEMATICA. Matrices for testing were taken from Ref. [88], which is observed in the case of partial $a = 1$.

First row of Table 3.1 contains the names of test matrices from Ref. [88]. Note that the test three groups of test matrices. The last row of the table contains the time obtained in the LDLGInverse algorithm. Symbol '-' denotes excessive time of calculating.

Example 3.2.3. Consider now randomly generated rare possessed matrix from the set $A \in \mathbf{R}_r^{m\times n}$ different sizes and densities, one of the largest eigenvalue of the matrix A^*A given with k_r, a minimum non-zero net value matrix A^*A and it is equal to 1 (see [12]). Some of the results obtained are shown in Table 3.2. Table 3.3 shows the difference between the execution time of algorithms for some matrix from collection Matrix-Market (see [39]).

It was noted that some very quick method numerically unstable if the die (singular or not) lose conditioned (see [12, 79]). Introduced method for calculating the generalized inverse [58]is very efficient, however, is not the most efficient. Obviously, the rank-deficient matrix processed faster than full-rank matrix of the same dimensions, since the matrix L and D smaller, in the case of rank-deficient matrix. However, the calculation time is rapidly growing in size and increasing the density matrix. In the general case, the calculation of the inverse matrix is rational $O(n^3)$ problem, but also very sensitive to bad conditioned matrices.

It is well-known that the conditional code for matrix inversion compared to standard dot matrix $\|\cdot\|$ square matrix A, defined as $\kappa(A) = \|A\| \|A^{-1}\|$, a measure of stability and sensitivity of inverse matrix [19]. Matrix with the conditioning number close 1 are called well-conditioned, and those with greater strength and conditioning number of one called lose-conditioned.

Table 3.1. Execution Time (in seconds) Obtained by Several Algorithms and LDLGInverse Algorithm

Test matrix	A10	A50	A100	S10	S50	S100	F10	F50	F100
PseudoInverse [86]	1.124	-	-	1.001	-	-	1.022	-	-
Partitioning [65]	0.012	0.284	1.650	0.02	1.084	6.204	0.016	0.445	2.812
Lev.-Faddeev [29]	0.001	1.78	41.581	0.001	2.221	42.112	0.001	1.872	40.844
Courrieu [11]	0.012	0.664	4.564	0.013	0.225	2.197	0.015	0.622	5.552
ModCholesky [69]	0.013	2.2	16.22	0.014	0.63	5.75	0.014	2.1	16.22
LDLGInverse	0.014	1.35	10.44	0.013	0.399	4.44	0.014	1.782	11.2

Table 3.2. Average Time of Calculation (in seconds) for the Randomly Generated Rare Possessed Matrix

m	128			256			512		
k_r	16	256	4096	16	256	4096	16	256	4096
Full-rank	4.22	4.23	4.33	29.4	30.2	30.3	222.33	223.11	234.55
Rank-deficient	3.01	3.23	3.44	22.1	23.21	24.57	182.3	183.24	185.46

Table 3.3. Average Time of Calculation (in seconds) for Some Test Matrices from Collection Matrix-Market

	gr 30 30	illc1850	watt 1	well1033	well1850
Matrix Dimensions	900900	1850712	18561856	1033320	1850712
Matrix Density	0.0096	0.0066	0.0033	0.0143	0.0066
Timings	221.7	122.1	38.4	10.6	114.23

Note that all five matrices taken from "MatrixMarket" tested in Table 3.3 are conditional numbers greater than one, whereby `MATHEMATICA` displays the information "*results for inverse of badly conditioned matrix may contain significant numerical errors.*" LDLGInverse algorithm is designed primarily for symbolic calculations with polynomial and rational matrices where it is desirable avoid the square roots.

3.3 LDL^* Full-Rank Decomposition of Polynomial Matrix

As noted, the replacement LL^* factorization with LDL^*, so as to avoid calculating the elements with square roots, is of great importance in dealing with polynomial and rational matrices. Thus, the motivation is modified Cholesky method decomposition polynomial matrix $A(x)$ by introducing additional diagonal D matrix that provides non-occurrence of elements containing square roots. Obviously, LDL^* decomposition is much more convenient to work with polynomial matrices, and later can be exploited for finding generalized inverse matrix factored.

Recall that in the case of a complex Hermitian matrix A, it is possible to determine its LDL^* decomposition using recursive relations (3.2.1). In papers [58] and [72] we proposed that the calculation (3.2.1) is necessary to execute only $j = \overline{1,r}$, where $r = \text{rank}(A)$. In this case, iterative procedure (3.2.1) generates factorization full-rank of matrix A, where the matrix L

will be without zero root, and matrix D types without a zero column and zero (with respect to the matrix L and D then also rank r).

Consider the polynomial Hermitian matrix $A \in \mathbf{C}(x)_r^{n\times n}$ rank r with elements:

$$a_{ij}(x) = \sum_{k=0}^{a_q} a_{k,i,j}x^k, \quad 1 \le i,\ j \le n, \tag{3.3.1}$$

whereby the maximum exponent of the matrix $A(x)$ labeled with a_q. Then the free-roots Cholesky decomposition of the full-rank of the matrix A is given by $A = LDL^*$, wherein $L \in \mathbf{C}(x)^{n\times r}$, $l_{ij} = 0$ for $i < j$, a $D \in \mathbf{C}(x)^{r\times r}$ the corresponding diagonal matrix rational. The non-zero elements of rational matrix $L(x)$ and $D(x)$ shape:

$$\begin{aligned} d_{jj}(x) &= \frac{\sum\limits_{k=0}^{\overline{d}_q} \overline{d}_{k,j,j}x^k}{\sum\limits_{k=0}^{\overline{\overline{d}}_q} \overline{\overline{d}}_{k,j,j}x^k}, \quad l_{jj}(x) = 1, \quad 1 \le j \le r, \\ l_{ij}(x) &= \frac{\sum\limits_{k=0}^{\overline{l}_q} \overline{l}_{k,i,j}x^k}{\sum\limits_{k=0}^{\overline{\overline{l}}_q} \overline{\overline{l}}_{k,i,j}x^k}, \quad 1 \le j \le n, \quad 1 \le i \le r, \quad j < i. \end{aligned} \tag{3.3.2}$$

Symbolic general calculation of inverses and their gradients is one of the most interesting areas of computer algebra. Greville's method of alterations algorithm are commonly used in a symbolic implementations of generalized inverse. Several extensions of the algorithm partitions papers rational and polynomial matrix is introduced. The first is an extension and generalization of Greville's algorithm on a set of polynomial and/or rational matrix of an unknown, introduced in Ref. [65]. Further implementation and modification of algorithms for calculating weighting Moore–Penrose inverse of programs can be found in the Refs. [63, 74]. For more information about calculating Drazin inverses, Moore–Penrose's inverse weighting and Moore–Penrose's inverse, see [14, 15, 16, 64, 68].

Motivation is perfected an efficient method for calculating symbolic Moore–Penrose's inverse polynomial matrix. By using the LDL^* Cholesky factorization rather than classical decomposition, avoided obtaining elements with square roots, which is of essential importance in calculating symbolic polynomial. For symbolic calculation Moore–Penrose's inverse is applied Theorem 3.2.7, introduced in Ref. [58], the polynomial matrices that occur during the calculation.

Already mentioned iterative procedure (3.2.1) can be modified so that the products as a result of decomposition of the full-rank of the matrix polynomial A (x). Accordingly, the following relation vases for rational elements of the matrix $D(x)$ and $L(x)$ for each $j = \overline{1,r}$:

$$f_{ij}(x) = \sum_{k=1}^{j-1} l_{ik}(x)l^*_{jk}(x)d_{kk}(x), \text{ for } i = \overline{j,n}, \tag{3.3.3}$$

$$d_{jj}(x) = a_{jj}(x) - f_{jj}(x), \tag{3.3.4}$$

$$l_{ij}(x) = \frac{1}{d_{jj}(x)}(a_{ij}(x) - f_{ij}(x)), \text{ for } i = \overline{j+1,n}. \tag{3.3.5}$$

These recurrent relations are used in the following theorem for calculating the coefficients of rational matrix L and D. Below dissertation variable with one line will denote coefficients in the numerator, while the variables with two lines denote coefficients in the denominator fraction.

Theorem 3.3.1. [72] *LDL^* decomposition of the full-rank Hermitian polynomial matrix $A(x) \in \mathbf{C}(x)^{n\times n}_r$ with the elements form* (3.3.1) *is equal $A(x) = L(x)D(x)L(x)^*$, where $L(x)$ and $D(x)$ rational matrix in form* (3.3.2)*, whereby the coefficients of elements $d_{jj}(x)$ and $l_{ij}(x)$ given the fol-*

lowing terms:

$$\overline{d}_{k,j} = \sum_{k_1=0}^{k} a_{k-k_1,j,j}\overline{\overline{f}}_{k_1,j,j} - \overline{f}_{k,j,j},\ 0 \le k \le \overline{d}_q = \quad (3.3.6)$$

$$\max(a_q + \overline{\overline{f}}_q, \overline{f}_q), \quad (3.3.7)$$

$$\overline{\overline{d}}_{k,j} = \overline{\overline{f}}_{k,j,j},\ 0 \le k \le \overline{\overline{d}}_q = \overline{\overline{f}}_q, \quad (3.3.8)$$

$$\overline{l}_{k,i,j} = \sum_{k_1=0}^{k} \overline{\overline{d}}_{k-k_1,j}\left(\sum_{k_2=0}^{k_1} a_{k_1-k_2,i,j}\overline{\overline{f}}_{k_2,i,j} - \overline{f}_{k_1,i,j}\right),$$

$$0 \le k \le \overline{l}_q = \overline{\overline{d}}_q + \max(a_q + \overline{\overline{f}}_q, \overline{f}_q), \quad (3.3.9)$$

$$\overline{\overline{l}}_{k,i,j} = \sum_{k_1=0}^{k} \overline{d}_{k-k_1,j}\overline{\overline{f}}_{k_1,i,j},\ 0 \le k \le \overline{\overline{l}}_q = \overline{d}_q + \overline{\overline{f}}_q, \quad (3.3.10)$$

where are coefficients $\overline{f}_{k,i,j}$ *equal*

$$\overline{f}_{k,i,j} = \sum_{k_2=0}^{k}\sum_{k_3=0}^{j-1} \overline{p}_{k-k_2,i,j,k_3} q_{k_2,i,j,k_3}, \quad (3.3.11)$$

$$0 \le k \le \overline{f}_q = 2\overline{l}_q + \overline{d}_q + \overline{\overline{f}}_q - 2\overline{\overline{l}}_q - \overline{\overline{d}}_q, \quad (3.3.12)$$

a $\overline{\overline{f}}_{k,i,j}$, $0 \le k \le \overline{\overline{f}}_q$, *the coefficients of the polynomial next:*

$$LCM\left(\sum_{k=0}^{2\overline{\overline{l}}_q+\overline{\overline{d}}_q} \overline{\overline{p}}_{k,i,j,1}x^k, \sum_{k=0}^{2\overline{\overline{l}}_q+\overline{\overline{d}}_q} \overline{\overline{p}}_{k,i,j,2}x^k, \ldots, \sum_{k=0}^{2\overline{\overline{l}}_q+\overline{\overline{d}}_q} \overline{\overline{p}}_{k,i,j,j-1}x^k\right), \quad (3.3.13)$$

wherein

$$\overline{p}_{t_1,i,j,k} = \sum_{t_2=0}^{t_1}\sum_{t_3=0}^{t_1-t_2} \overline{l}_{t_3,i,k}\overline{l}^*_{t_1-t_2-t_3,j,k}\overline{d}_{t_2,k},\ 0 \le t_1 \le 2\overline{l}_q + \overline{d}_q,$$

$$\overline{\overline{p}}_{t_1,i,j,k} = \sum_{t_2=0}^{t_1}\sum_{t_3=0}^{t_1-t_2} \overline{\overline{l}}_{t_3,i,k}\overline{\overline{l}}^*_{t_1-t_2-t_3,j,k}\overline{\overline{d}}_{t_2,k},\ 0 \le t_1 \le 2\overline{\overline{l}}_q + \overline{\overline{d}}_q,$$

while values $q_{k,i,j,t}$ *are coefficients of polynomials.* $q_{i,j,t}(x) = \dfrac{\sum_{k=0}^{\overline{\overline{f}}_q} \overline{\overline{f}}_{k,i,j}x^k}{\sum_{k=0}^{2\overline{\overline{l}}_q+\overline{\overline{d}}_q} \overline{\overline{p}}_{k,i,j,t}x^k}$.

Proof. How are the elements of the matrix $L(x)$ and $D(x)$ of rational functions, equations (3.3.3) received the following form:

$$
\begin{aligned}
f_{ij}(x) &= \\
&= \sum_{k=1}^{j-1} \frac{\sum_{t=0}^{\overline{l}_q} \overline{l}_{t,i,k} x^t}{\sum_{t=0}^{\overline{\overline{l}}_q} \overline{\overline{l}}_{t,i,k} x^t} \frac{\sum_{t=0}^{\overline{l}_q} \overline{l}^*_{t,j,k} x^t}{\sum_{t=0}^{\overline{\overline{l}}_q} \overline{\overline{l}}^*_{t,j,k} x^t} \frac{\sum_{t=0}^{\overline{d}_q} \overline{d}_{t,k} x^t}{\sum_{t=0}^{\overline{\overline{d}}_q} \overline{\overline{d}}_{t,k} x^t} \\
&= \sum_{k=1}^{j-1} \frac{\sum_{t_1=0}^{2\overline{l}_q+\overline{d}_q} \left(\sum_{t_2=0}^{t_1} \sum_{t_3=0}^{t_1-t_2} \overline{l}_{t_3,i,k} \overline{l}^*_{t_1-t_2-t_3,j,k} \overline{d}_{t_2,k} \right) x^{t_1}}{\sum_{t_1=0}^{2\overline{\overline{l}}_q+\overline{\overline{d}}_q} \left(\sum_{t_2=0}^{t_1} \sum_{t_3=0}^{t_1-t_2} \overline{\overline{l}}_{t_3,i,k} \overline{\overline{l}}^*_{t_1-t_2-t_3,j,k} \overline{\overline{d}}_{t_2,k} \right) x^{t_1}} \\
&= \sum_{k=1}^{j-1} \frac{\sum_{t_1=0}^{2\overline{l}_q+\overline{d}_q} \overline{p}_{t_1,i,j,k} x^{t_1}}{\sum_{t_1=0}^{2\overline{\overline{l}}_q+\overline{\overline{d}}_q} \overline{\overline{p}}_{t_1,i,j,k} x^{t_1}} = \frac{\sum_{t_1=0}^{2\overline{l}_q+\overline{d}_q} \overline{p}_{t_1,i,j,1} x^{t_1}}{\sum_{t_1=0}^{2\overline{\overline{l}}_q+\overline{\overline{d}}_q} \overline{\overline{p}}_{t_1,i,j,1} x^{t_1}} + \frac{\sum_{t_1=0}^{2\overline{l}_q+\overline{d}_q} \overline{p}_{t_1,i,j,2} x^{t_1}}{\sum_{t_1=0}^{2\overline{\overline{l}}_q+\overline{\overline{d}}_q} \overline{\overline{p}}_{t_1,i,j,2} x^{t_1}} + \\
&\quad \ldots + \frac{\sum_{t_1=0}^{2\overline{l}_q+\overline{d}_q} \overline{p}_{t_1,i,j,j-1} x^{t_1}}{\sum_{t_1=0}^{2\overline{\overline{l}}_q+\overline{\overline{d}}_q} \overline{\overline{p}}_{t_1,i,j,j-1} x^{t_1}}.
\end{aligned}
$$

Given that the lowest common denominator (LCM) in the denominator polynomial labeled as:

$$
LCM\left(\sum_{k=0}^{2\overline{\overline{l}}_q+\overline{\overline{d}}_q} \overline{\overline{p}}_{k,i,j,1} x^k, \sum_{k=0}^{2\overline{\overline{l}}_q+\overline{\overline{d}}_q} \overline{\overline{p}}_{k,i,j,2} x^k, \ldots, \sum_{k=0}^{2\overline{\overline{l}}_q+\overline{\overline{d}}_q} \overline{\overline{p}}_{k,i,j,j-1} x^k \right)
$$
$$
= \sum_{k=0}^{\overline{\overline{f}}_q} \overline{\overline{f}}_{k,i,j} x^k,
$$

and vases of equality

$$q_{i,j,t}(x) = \frac{\sum\limits_{k=0}^{\overline{\overline{f}}_q} \overline{\overline{f}}_{k,i,j} x^k}{\sum\limits_{k=0}^{2\overline{\overline{l}}_q+\overline{\overline{d}}_q} \overline{\overline{p}}_{k,i,j,t} x^k} = \sum_{k=0}^{\overline{\overline{f}}_q-2\overline{\overline{l}}_q-\overline{\overline{d}}_q} q_{k,i,j,t} x^k, \quad 1 \le t < j < i,$$

polynomials $f_{ij}(x)$ can be expressed as

$$\begin{aligned} f_{ij}(x) &= \frac{\sum\limits_{k=1}^{j-1} \sum\limits_{k_1=0}^{\overline{f}_q} \left(\sum\limits_{k_2=0}^{k_1} \overline{p}_{k_1-k_2,i,j,k} q_{k_2,i,j,k} \right) x^{k_1}}{\sum\limits_{k=0}^{\overline{\overline{f}}_q} \overline{\overline{f}}_{k,i,j} x^k} \\ &= \frac{\sum\limits_{k_1=0}^{\overline{f}_q} \sum\limits_{k_2=0}^{k_1} \sum\limits_{k_3=0}^{j-1} \overline{p}_{k_1-k_2,i,j,k_3} q_{k_2,i,j,k_3} x^{k_1}}{\sum\limits_{k=0}^{\overline{\overline{f}}_q} \overline{\overline{f}}_{k,i,j} x^k} = \frac{\sum\limits_{k=0}^{\overline{f}_q} \overline{f}_{k,i,j} x^k}{\sum\limits_{k=0}^{\overline{\overline{f}}_q} \overline{\overline{f}}_{k,i,j} x^k}. \end{aligned}$$

Note that from the equation (3.3.4) followed by the next equation:

$$\begin{aligned} d_{jj}(x) &= \sum_{k=0}^{a_q} a_{k,j,j} x^k - \frac{\sum\limits_{k=0}^{\overline{f}_q} \overline{f}_{k,j,j} x^k}{\sum\limits_{k=0}^{\overline{\overline{f}}_q} \overline{\overline{f}}_{k,j,j} x^k} \\ &= \frac{\sum\limits_{k=0}^{a_q} a_{k,j,j} x^k \sum\limits_{k=0}^{\overline{\overline{f}}_q} \overline{\overline{f}}_{k,j,j} x^k - \sum\limits_{k=0}^{\overline{f}_q} \overline{f}_{k,j,j} x^k}{\sum\limits_{k=0}^{\overline{\overline{f}}_q} \overline{\overline{f}}_{k,j,j} x^k} \end{aligned}$$

$$= \frac{\sum_{k_1=0}^{\max(a_q+\overline{\overline{f}}_q,\overline{f}_q)} \left(\sum_{k_2=0}^{k_1} a_{k_1-k_2,j,j} \overline{\overline{f}}_{k_2,j,j} - \overline{f}_{k_1,j,j} \right) x^{k_1}}{\sum_{k=0}^{\overline{\overline{f}}_q} \overline{\overline{f}}_{k,j,j} x^k}$$

$$= \frac{\sum_{k=0}^{\overline{d}_q} \overline{d}_{k,j} x^k}{\sum_{k=0}^{\overline{\overline{d}}_q} \overline{\overline{d}}_{k,j} x^k}.$$

Finally, on the basis of Eq. (3.3.5), following equality is valid:

$$l_{ij}(x) = \frac{\sum_{k=0}^{\overline{\overline{d}}_q} \overline{\overline{d}}_{k,j} x^k}{\sum_{k=0}^{\overline{d}_q} \overline{d}_{k,j} x^k} \left(\sum_{k=0}^{a_q} a_{k,i,j} x^k - \frac{\sum_{k=0}^{\overline{f}_q} \overline{f}_{k,i,j} x^k}{\sum_{k=0}^{\overline{\overline{f}}_q} \overline{\overline{f}}_{k,i,j} x^k} \right)$$

$$= \frac{\sum_{k=0}^{\overline{\overline{d}}_q} \overline{\overline{d}}_{k,j} x^k}{\sum_{k=0}^{\overline{d}_q} \overline{d}_{k,j} x^k} \frac{\sum_{k_1=0}^{\max(a_q+\overline{\overline{f}}_q,\overline{f}_q)} \left(\sum_{k_2=0}^{k_1} a_{k_1-k_2,i,j} \overline{\overline{f}}_{k_2,i,j} - \overline{f}_{k_1,i,j} \right) x^{k_1}}{\sum_{k=0}^{\overline{\overline{f}}_q} \overline{\overline{f}}_{k,j,j} x^k}$$

$$= \frac{\sum_{k=0}^{\overline{\overline{d}}_q+\max(a_q+\overline{\overline{f}}_q,\overline{f}_q)} \left(\sum_{k_1=0}^{k} \overline{\overline{d}}_{k-k_1,j} \left(\sum_{k_2=0}^{k_1} a_{k_1-k_2,i,j} \overline{\overline{f}}_{k_2,i,j} - \overline{f}_{k_1,i,j} \right) \right) x^k}{\sum_{k=0}^{\overline{d}_q+\overline{\overline{f}}_q} \left(\sum_{k_1=0}^{k} \overline{d}_{k-k_1,j} \overline{\overline{f}}_{k_1,i,j} \right) x^k}$$

$$= \frac{\sum_{k=0}^{\overline{l}_q} \overline{l}_{k,i,j} x^k}{\sum_{k=0}^{\overline{\overline{l}}_q} \overline{\overline{l}}_{k,i,j} x^k}.$$

□

The previous theorem provides a practical method for calculating the coefficients appearing in the elements $l_{ij}(x)$ and $d_{jj}(x)$ based on previously the calculated $l_{ik}(x)$, $l_{jk}(x)$ and $d_{kk}(x)$, for $k < j$. Now the algorithm to calculate LDL^* complete decomposition rank date Hermitian polynomial matrix $A(x) = \{a_{ij}(x)\}_{i,j=1}^{n}$ can be formulated.

Algorithm 3.2 Calculating LDL^* decomposition of the full-rank given Hermitian polynomial matrix

---

Input: Hermitian polynomial matrix $A(x)$ rank r.

1: Initialization: $d_1(x) := a_{11}(x)$, $l_{i1}(x) := \frac{a_{i1}(x)}{a_{11}(x)}$ for $i = \overline{2,n}$. Set the diagonal elements of the matrix $L(x)$ to 1. For $j = \overline{2,r}$ repeating steps 2, 3, 4.

2: Calculate the coefficients of the polynomial $f_{ij}(x)$ through the following steps. For $i = \overline{j,n}$ repeat steps 2.1, 2.2, 2.3, and 2.4.

2.1: For each $k = \overline{1, j-1}$ calculate

$$\overline{p}_{t_1,i,j,k} = \sum_{t_2=0}^{t_1}\sum_{t_3=0}^{t_1-t_2} \overline{l}_{t_3,i,k}\overline{l}^*_{t_1-t_2-t_3,j,k}\overline{d}_{t_2,k}, \; 0 \le t_1 \le 2\overline{l}_q + \overline{d}_q,$$

$$\overline{\overline{p}}_{t_1,i,j,k} = \sum_{t_2=0}^{t_1}\sum_{t_3=0}^{t_1-t_2} \overline{\overline{l}}_{t_3,i,k}\overline{\overline{l}}^*_{t_1-t_2-t_3,j,k}\overline{\overline{d}}_{t_2,k}, \; 0 \le t_1 \le 2\overline{\overline{l}}_q + \overline{\overline{d}}_q.$$

2.2: determine polynomial smallest together for joint denominator polynomial generated determining the denominators:

$$\sum_{k=0}^{\overline{\overline{f}}_q} \overline{\overline{f}}_{k,i,j}x^k = LCM\left(\sum_{k=0}^{2\overline{\overline{l}}_q+\overline{\overline{d}}_q} \overline{\overline{p}}_{k,i,j,1}x^k, \ldots, \sum_{k=0}^{2\overline{\overline{l}}_q+\overline{\overline{d}}_q} \overline{\overline{p}}_{k,i,j,j-1}x^k\right).$$

2.3: For each $t = \overline{1, j-1}$ and each $i > j$ divide polynomial $\sum_{k=0}^{\overline{\overline{f}}_q} \overline{\overline{f}}_{k,i,j}x^k$ with the following polynomial: $\sum_{k=0}^{2\overline{\overline{l}}_q+\overline{\overline{d}}_q} \overline{\overline{p}}_{k,i,j,t}x^k$, and denote quotient

$$q_{i,j,t}(x) = \sum_{k=0}^{\overline{\overline{f}}_q - 2\overline{\overline{l}}_q - \overline{\overline{d}}_q} q_{k,i,j,t}x^k.$$

2.4: Introduce the substitute $\overline{f}_q = 2\overline{l}_q + \overline{d}_q + \overline{\overline{f}}_q - 2\overline{\overline{l}}_q - \overline{\overline{d}}_q$, and calculate

$$\overline{f}_{k,i,j} = \sum_{k_2=0}^{k} \sum_{k_3=0}^{j-1} \overline{p}_{k-k_2,i,j,k_3} q_{k_2,i,j,k_3}, \ 0 \le k \le \overline{f}_q.$$

3: Introduce the substitute $\overline{d}_q = \max(a_q + \overline{\overline{f}}_q, \overline{f}_q)$ and $\overline{\overline{d}}_q = \overline{\overline{f}}_q$. Make the following evaluation:

$$\begin{aligned} \overline{d}_{k,j} &= \sum_{i=0}^{k} a_{k-i,j,j} \overline{\overline{f}}_{i,j,j} - \overline{f}_{k,j,j}, \ 0 \le k \le \overline{d}_q \\ \overline{\overline{d}}_{k,j} &= \overline{\overline{f}}_{k,j,j}, \ 0 \le k \le \overline{\overline{d}}_q. \end{aligned}$$

4: Introduce the substitution $\overline{l}_q = \overline{\overline{d}}_q + \max(a_q + \overline{\overline{f}}_q, \overline{f}_q)$ i $\overline{\overline{l}}_q = \overline{d}_q + \overline{\overline{f}}_q$. For each $i = \overline{j+1,n}$ make the following evaluation

$$\begin{aligned} \overline{l}_{k,i,j} &= \sum_{k_1=0}^{k} \overline{\overline{d}}_{k-k_1,j} \left(\sum_{k_2=0}^{k_1} a_{k_1-k_2,i,j} \overline{\overline{f}}_{k_2,i,j} - \overline{f}_{k_1,i,j} \right), \\ \overline{\overline{l}}_{k,i,j} &= \sum_{k_1=0}^{k} \overline{d}_{k-k_1,j} \overline{\overline{f}}_{k_1,i,j}, \ 0 \le k \le \overline{\overline{l}}_q. \end{aligned}$$

5: **Rezultat:**

$$d_j(x) = \frac{\sum_{k=0}^{\overline{d}_q} \overline{d}_{k,j} x^k}{\sum_{k=0}^{\overline{\overline{d}}_q} \overline{\overline{d}}_{k,j} x^k}, \quad l_{i,j}(x) = \frac{\sum_{k=0}^{\overline{l}_q} \overline{l}_{k,i,j} x^k}{\sum_{k=0}^{\overline{\overline{l}}_q} \overline{\overline{l}}_{k,i,j} x^k},$$

$$j = \overline{1,r}, \ i = \overline{j+1,n}.$$

Here above algorithm applies only multiplication, division and summarizing matrix elements. In doing so, this method does not involve the use of square root, which as a consequence means creating a new diagonal matrix. Obvious, This algorithm is very suitable for implementation in procedural programming languages.

3.3.1 Numerical Examples

Example 3.3.1. Consider a symmetrical polynomial matrix rank 2 generated in the work [88]:

$$S_3 = \begin{bmatrix} 1+x & x & 1+x \\ x & -1+x & x \\ 1+x & x & 1+x \end{bmatrix}.$$

For $j = 1$ it is $d_{11} = 1+x$, hence

$$l_{21}(x) = \frac{x}{1+x}, \quad l_{31}(x) = \frac{1+x}{1+x} = 1.$$

For case $j = 2$ we have that

$$f_{22}(x) = \frac{x^2}{1+x}, \quad f_{32}(x) = x.$$

Based on these results is satisfied:

$$d_{22}(x) = -\frac{1}{1+x}, \quad l_{32}(x) = \frac{1}{d_{22}(x)}(x - f_{32}(x)) = 0,$$

hence they receive the following rational matrix:

$$L(x) = \begin{bmatrix} 1 & 0 \\ \frac{x}{1+x} & 1 \\ 1 & 0 \end{bmatrix}, \quad D(x) = \begin{bmatrix} 1+x & 0 \\ 0 & -\frac{1}{1+x} \end{bmatrix}.$$

Example 3.3.2. Consider the symmetric matrix polynomial

$$A(x) = \begin{bmatrix} 1+x^2 & x & 2+x & -3+x \\ x & -1+x & 3+x & x \\ 2+x & 3+x & -1+x & 1+x \\ -3+x & x & 1+x & x \end{bmatrix}$$

In Step 1 of Algorithm 3.2 we will put $d_1(x) = 1+x^2$ and

$$l_{21}(x) = \frac{x}{1+x^2}, \; l_{31}(x) = \frac{2+x}{1+x^2}, \; l_{41}(x) = \frac{-3+x}{1+x^2}.$$

For $j = 2$ we have that

$$\mathrm{f}_{22}(x)=\frac{x^2}{1+x^2},\ \mathrm{f}_{32}(x)=\frac{x(2+x)}{1+x^2},\ \mathrm{f}_{42}(x)=\frac{(-3+x)x}{1+x^2}.$$

Therefore, the simple is obtained

$$d_{22}(x) = (-1+x-2x^2+x^3)/(1+x^2)$$

a non-zero elements of the matrix L are

$$l_{32}(x) = \frac{3-x+2x^2+x^3}{-1+x-2x^2+x^3},\ l_{42}(x) = \frac{x(4-x+x^2)}{-1+x-2x^2+x^3}.$$

For $j = 3$ is gained

$$\mathrm{f}_{33}(x)=\frac{5-6x+3x^2+5x^3+x^4}{-1+x-2x^2+x^3},\ \mathrm{f}_{43}(x) = \frac{6+7x-3x^2+2x^3+x^4}{-1+x-2x^2+x^3}$$

therefore follows

$$d_{33}(x) = \frac{-4+4x-8x^3}{-1+x-2x^2+x^3},\ l_{43}(x) = \frac{7+7x-2x^2+3x^3}{4-4x+8x^3}.$$

Finally, for $j = 4$ is gained $f_{44}(x) = (85+87x-2x^2-x^3+8x^4)/(4-4x+8x^3)$.This result leads to the

$$d_{44}(x) = \frac{-85-83x-2x^2+x^3}{4-4x+8x^3}.$$

Thus, we have obtained from the following matrix factorization:

$$L(x) = \begin{bmatrix} 1 & 0 & 0 & 0 \\ \frac{x}{1+x^2} & 1 & 0 & 0 \\ \frac{2+x}{1+x^2} & \frac{3-x+2x^2+x^3}{-1+x-2x^2+x^3} & 1 & 0 \\ \frac{-3+x}{1+x^2} & \frac{x(4-x+x^2)}{-1+x-2x^2+x^3} & \frac{7+7x-2x^2+3x^3}{4-4x+8x^3} & 1 \end{bmatrix},$$

$$D(x) = \begin{bmatrix} 1+x^2 & 0 & 0 & 0 \\ 0 & \frac{-1+x-2x^2+x^3}{1+x^2} & 0 & 0 \\ 0 & 0 & \frac{-4+4x-8x^3}{-1+x-2x^2+x^3} & 0 \\ 0 & 0 & 0 & \frac{-85-83x-2x^2+x^3}{4-4x+8x^3} \end{bmatrix}.$$

3.3.2 Implementation Details and the Results of Testing

Since the LDL^* decomposition of the same complexity as well as Cholesky factorization, the efficiency of the algorithm will not be different from the efficiency of the algorithm LL^* decomposition. Algorithm 3.2 means one more multiplying matrix and the diagonal matrix. Thus, the computational complexity will be on the scale $O(n^3m^2)$, wherein $O(n^3)$ complexity LDL^* decomposition, a m is the maximum exponent of the polynomial obtained in the algorithm.

Obviously sometimes coefficients of the polynomial can be a growth in inter-steps, even though simplification is executed. One possible solutions to this problem is to consider implementation large number operations. Another possibility is to handle this with matrix elements being quotients two polynomials, thereat doing simplification after each step. Polynomials matrices containing a relative small number non-zero entries are often considered in practical computations. In this case, previous algorithm is not very effective because many redundant operations.

Procedure `PolynomialLDLDecomposition[A_List]` for testing and check the steps of the algorithm in the package `MATHEMATICA`, is given in Ref. [59], and here is the data as a whole.

```
PolynomialLDLDecomposition[A_List] :=
 Module[{i, j, k, n = Length[A], r = MatrixRank[A], L, D,
   f, p, q, Num, Den, DNum, LNum, LDen, UpperLim = 0, x, r1},
  L = f = Table[0, {n}, {r}];
  D = Table[0, {r}, {r}];
  p = q = Table[0, {3 n}, {3 n}];
  D[[1, 1]] = A[[1, 1]];
  For[j = 1, j ≤ r, j++, L[[j, j]] = 1];
  For[i = 2, i ≤ n, i++, L[[i, 1]] = Simplify[A[[i, 1]] / A[[1, 1]]]];
  For[j = 2, j ≤ r, j++,
   For[i = j, i ≤ n, i++,
     For[k = 1, k ≤ j - 1, k++,
      Num[k] = 0; Den[k] = 0;
      For[r1 = 0, r1 ≤ 2 * Exponent[L[[i, k]], x] + Exponent[D[k, k], x], r1++,
       p[[r1 + 1, k]] = 0; q[[r1 + 1, k]] = 0;
       p[[r1 + 1, k]] = ∑_{r2=0}^{r1} ∑_{r3=0}^{r1-r2} Coefficient[Numerator[L[[i, k]]], x, r3] *
            Coefficient[Numerator[L[[j, k]]], x, r1 - r2 - r3] *
            Coefficient[Numerator[D[[k, k]]], x, r2];
       q[[r1 + 1, k]] = ∑_{r2=0}^{r1} ∑_{r3=0}^{r1-r2} Coefficient[Denominator[L[[i, k]]], x, r3] *
            Coefficient[Denominator[L[[j, k]]], x, r1 - r2 - r3] *
            Coefficient[Denominator[D[[k, k]]], x, r2];
       Num[k] += p[[r1 + 1, k]] * x^r1;
       Den[k] += q[[r1 + 1, k]] * x^r1;
      ];
     ];
     f[[i, j]] = Simplify[∑_{k=1}^{j-1} Num[k] / Den[k]];
     Print["f[", i, j, "]=", f[[i, j]]];
   ];
```

```
DNum = 0;
  UpperLim = Max[Max[Exponent[A, x]] + Max[Exponent[Denominator[f[[j, j]]], x]],
    Max[Exponent[Numerator[f[[j, j]]], x]]];
  For[k = 0, k ≤ UpperLim, k++,
   DNum += ((Sum[Coefficient[A[[j, j]], x, k - i] *
          Coefficient[Denominator[f[[j, j]]], x, i], {i, 0, k}]) -
        Coefficient[Numerator[f[[j, j]]], x, k]) * x^k];
  D[[j, j]] = Simplify[DNum / Denominator[f[[j, j]]]];
  For[i = j + 1, i ≤ n, i++,
   LNum = 0;
   For[k = 0, k ≤ Max[Exponent[Denominator[f[[i, j]]], x]] + UpperLim, k++,
    LNum += Sum[Coefficient[Denominator[D[[j, j]]], x, k - k1] *
        ((Sum[Coefficient[A[[i, j]], x, k1 - k2] * Coefficient[Denominator[f[[i, j]]],
              x, k2], {k2, 0, k1}]) - Coefficient[Numerator[f[[i, j]]], x, k1]) * x^k;, {k1, 0, k}]];
   LDen = 0;
   For[k = 0, k ≤ Max[Exponent[Denominator[f[[i, j]]], x]] + UpperLim, k++,
    LDen += (Sum[Coefficient[Numerator[D[[j, j]]], x, k - k1] *
          Coefficient[Denominator[f[[i, j]]], x, k1], {k1, 0, k}]) * x^k;];
   L[[i, j]] = Simplify[LNum / LDen];
   ];];
 Return[{Simplify[L], Simplify[D]}];]
```

It is designed and package `PolynomialMatrix` in `Java` for symbolic operations on polynomial matrices with elements from a set of $Z[x]$. This package contains four main classes: Polynomial, Rational, Function, and Matrix and Decompositions. Polynomial implementation is implemented in by using a series of coefficients. Table of non-zero coefficients of the polynomial at the few. Even if the input and output polynomial algorithms in a very short odds in-between steps can be huge.

Therefore, were used two types of odds: long and BigInteger (Package for integer operations JAVA). Thus, we have obtained four classes, whereby we always use the most appropriate. The four basic operations are supported *Karatsuba* algorithm for polynomial multiplication and division to determine residues.

Use at high speed modular *NZD* algorithm for determining the greatest common divisor of two polynomials (By applying the Chinese Remainder Theorem, Euclid's algorithm). Polynomial $P(x)$ is free if all its coefficients are relatively prime. Coefficients and partial results grow rapidly in Euclidean algorithm with the polynomial sequence remains. Therefore, it is possible to calculate the free part the remainder in each step. However, a simple calculation works requires the calculation of the greatest common divisors of the polynomial whose coefficients can be large. Finally, we used the modular approach of the work [85], and using Landay-Mignotte border, each coefficient gcd polynomial with integer $a(x) = \sum_{i=0}^{m} a_i x^i$ and $b(x) = \sum_{i=0}^{n} b_i x^i$, where $a_m \neq 0$ and $b_n \neq 0$, is limited by the absolute value of the

$$2^{\min(m,n)} \cdot nzd(a_m, b_n) \cdot \min\left(\frac{1}{a_m}\sqrt{\sum_{i=0}^{m} a_i^2}, \frac{1}{b_n}\sqrt{\sum_{i=0}^{n} b_i^2}\right).$$

Next lemma is fundamental for the implementation of the `JAVA`.

Lemma 3.3.2. *Let $a, b \in Z[x]$ be two given polynomials, and let p be prime number which does not divide leading coefficients of polynomial $a(x)$ and $b(x)$. Let $a_{(p)}$ and $b_{(p)}$ be pictures from a and b modulo p, respectively. Let $c = nzd(a,b)$ be over the Z. Then $st(nzd(a_{(p)}, b_{(p)})) \geq st(nzd(a,b))$. Moreover, if p does not divide the resultant a/c and b/c, then $nzd(a_{(p)}, b_{(p)}) = c$* mod *p*.

Of course, gcd polynomial over Z_p is determined only by the accuracy of the product Constant. Then we use the Chinese Remainder Theorem for reconstruction coefficients the greatest common divisor.

Rational functions are preserved as the numerator and the denominator doubles which are free polynomials. After each operation, executes the simplification of the gcd fractional dividing a numerator and denominator. How to work with matrices maximum dimensions of the order 100, we used a regular matrix multiplication complexity $O(n^3)$, and Gauss-Jordan's elimination to calculate inverse, rank matrices and determinants.

Example 3.3.3. Here we present comparative results in calculation of time watching the different runtime implementation of the Algorithm 3.2 in `MATHEMATICA` and `JAVA`. All tests were carried out on the system Intel Core 2 Duo T5800 2.0 GHz sa 2 GB RAM memory, the Windows operating system, using `JRE 1.6.0_24` and `MATHEMATICA 7.0`. Any time is obtained as the mean value of the 20 independent runtime.

The set of test matrix consisting of $n \times n$ matrix with random nonzero coefficients in the interval $[-10, 10]$ and power d. Here `MATHEMATICA` implementation of the Algorithm 3.2 was superior to other implementations. The main reason for the simplification of the road that are performed in `MATHEMATICA`. Implementation of the basic procedures (3.3.3)–(3.3.5) is marked as `MATH. basic algorithm` and obviously a lot slower than the implementation of the Algorithm 3.2, which is denoted as `MATH. Algorithm 3.2`. Table 3.4 contains the observed computation timings, obtained from several implementations.

3.4 Calculating the Moore–Penrose's Inverse Polynomial Matrix

Based on the previous theorems 3.3.1it is possible to perform the following result theorems 3.2.7.

Theorem 3.4.1. [72] *Let* $A \in \mathbf{C}(x)_r^{m\times n}$ *be polynomial matrix with elements of the form* (3.3.1). *Consider* LDL^* *factorization full-rank of ma-*

Table 3.4. Time of Calculations (in seconds) Obtained from Different Algorithms and Random Test Matrices

n	d	`Basic algorithm`	`Algorithm 3.2`
5	10	0.24	0.03
5	25	0.70	0.07
5	50	1.66	0.14
5	100	5.8	0.34
5	200	19.1	1.01
10	2	0.12	0.06
10	5	0.42	0.13
10	10	1.04	0.13
10	25	3.16	0.44
10	50	8.61	1.21
10	100	27.52	2.71

trix $(A^*A)^*(A^*A)$, where $L \in \mathbf{C}(x)^{n\times r}$ and $D \in \mathbf{C}(x)^{r\times r}$ forms a matrix with the elements of (3.3.2). We denote the elements of the inverse matrix $N = (L^*LDL^*L)^{-1} \in \mathbf{C}(x)^{r\times r}$ as

$$n_{ij}(x) = \frac{\sum\limits_{t=0}^{\overline{n}_q} \overline{n}_{t,k,l}x^t}{\sum\limits_{t=0}^{\overline{\overline{n}}_q} \overline{\overline{n}}_{t,k,l}x^t}.$$

Then, an arbitrary element Moore–Penrose's inverse matrix A can be calculated as $A^{\dagger}_{ij}(x) = \frac{\overline{\Gamma}_{ij}(x)}{\overline{\overline{\Gamma}}_i(x)}$, where

$$\overline{\Gamma}_{ij}(x) = \sum_{t=0}^{\overline{\Gamma}_q-\overline{b}_q+\overline{b}_q} \left(\sum_{\mu=1}^{n} \sum_{l=1}^{\min\{\mu,r\}} \sum_{k=1}^{\min\{i,r\}} \sum_{\lambda=1}^{n} \sum_{\kappa=1}^{m} \sum_{t_1=0}^{t} \overline{\beta}_{t_1,i,j,k,l,\mu,\kappa,\lambda}\, \gamma_{t-t_1,i,k,l,\mu} \right) x^t,$$

$$\overline{\overline{\Gamma}}_i(x) = LCM\left\{ \sum_{t=0}^{\overline{\overline{b}}_q} \overline{\overline{\beta}}_{t,i,k,l,\mu}x^t \,\middle|\, \mu = \overline{1,n}, k = \overline{1,\min\{i,r\}}, l = \overline{1,\min\{\mu,r\}} \right\} \tag{3.4.1}$$

wherein $\overline{\overline{\Gamma}}_q$ maximum exponent in polynomials $\overline{\overline{\Gamma}}_i(x)$, $1 \le i \le m$, of values $\gamma_{t,i,k,l,\mu}$, $0 \le t \le \overline{\overline{\Gamma}}_q - \overline{\overline{b}}_q$, coefficients polynomial $\Gamma_{i,k,l,\mu}(x) = \frac{\overline{\overline{\Gamma}}_i(x)}{\sum_{t=0}^{\overline{\overline{b}}_q} \overline{\overline{\beta}}_{t,i,k,l,\mu}x^t}$,

for each $\mu = \overline{1,n}$, $k = \overline{1,\min\{i,r\}}$, $l = \overline{1,\min\{\mu,r\}}$,
whereby the following notations were used for $\kappa = \overline{1,m}$, $\lambda = \overline{1,n}$:

$$\begin{aligned}
&\overline{\beta}_{t,i,j,k,l,\mu,\kappa,\lambda} = \sum_{t_1=0}^{t} \overline{p}_{t_1,i,k,l,\mu}\alpha_{t-t_1,j,\mu,\kappa,\lambda},\\
&0 \le t \le \overline{b}_q = 2\overline{l}_q + \overline{n}_q + 3\overline{a}_q,\\
&\overline{\overline{\beta}}_{t,i,k,l,\mu} = \sum_{t_1=0}^{t}\sum_{t_2=0}^{t-t_1} \overline{\overline{l}}_{t_1,i,k}\overline{\overline{n}}_{t-t_1-t_2,k,l}\overline{\overline{l}}^{*}_{t_2,\mu,l},\\
&0 \le t \le \overline{\overline{b}}_q = 2\overline{\overline{l}}_q + \overline{\overline{n}}_q,\\
&\overline{p}_{t,i,k,l,\mu} = \sum_{t_1=0}^{t}\sum_{t_2=0}^{t-t_1} \overline{l}_{t_1,i,k}\overline{n}_{t-t_1-t_2,k,l}\overline{l}^{*}_{t_2,\mu,l},\\
&0 \le t \le 2\overline{l}_q + \overline{n}_q\\
&\alpha_{t,j,\mu,\kappa,\lambda} = \sum_{t_1=0}^{t}\sum_{t_2=0}^{t-t_1} a_{t_1,\kappa,\lambda}a^{*}_{t-t_1-t_2,\kappa,\mu}a^{*}_{t_2,j,\lambda},\\
&0 \le t \le 3a_q.
\end{aligned} \tag{3.4.2}$$

Proof. Note that the following equality are met:

$$\begin{aligned}
(LNL^*)_{ij} &= \sum_{l=1}^{r}\left(\sum_{k=1}^{r} l_{ik}n_{kl}\right) l^*_{jl} = \sum_{l=1}^{r}\sum_{k=1}^{r} l_{ik}n_{kl}l^*_{jl},\\
((A^*A)^*A^*)_{ij} &= \sum_{l=1}^{n}\left(\sum_{k=1}^{m} a_{kl}a^*_{ki}\right) a^*_{jl} = \sum_{l=1}^{n}\sum_{k=1}^{m} a_{kl}a^*_{ki}a^*_{jl}.
\end{aligned}$$

Based on the first of the statements of Theorem 3.2.7, an arbitrary (i, j)-element called Moore–Penrose's inverse matrix A can be calculated as:

$$\begin{aligned}
A^{\dagger}_{ij} &= \sum_{\mu=1}^{n}\left(\sum_{l=1}^{\min\{\mu,r\}}\sum_{k=1}^{\min\{i,r\}} l_{ik}n_{kl}l^*_{\mu l}\right)\left(\sum_{\lambda=1}^{n}\sum_{\kappa=1}^{m} a_{\kappa\lambda}a^*_{\kappa\mu}a^*_{j\lambda}\right)\\
&= \sum_{\mu=1}^{n}\sum_{l=1}^{\min\{\mu,r\}}\sum_{k=1}^{\min\{i,r\}}\sum_{\lambda=1}^{n}\sum_{\kappa=1}^{m} l_{ik}n_{kl}l^*_{\mu l}a_{\kappa\lambda}a^*_{\kappa\mu}a^*_{j\lambda}.
\end{aligned}$$

Furthermore, the development of the polynomial in the previous term, (i,j)-ti element of matrix $A^{\dagger}$ can be calculated as:

$$A_{ij}^{\dagger}(x) = \sum_{\mu=1}^{n}\sum_{l=1}^{\min\{\mu,r\}}\sum_{k=1}^{\min\{i,r\}}\sum_{\lambda=1}^{n}\sum_{\kappa=1}^{m}\frac{\sum_{t=0}^{\overline{l}_q}\overline{l}_{t,i,k}x^t\sum_{t=0}^{\overline{n}_q}\overline{n}_{t,k,l}x^t\sum_{t=0}^{\overline{l}_q}\overline{l}^*_{t,\mu,l}x^t}{\sum_{t=0}^{\overline{\overline{l}}_q}\overline{\overline{l}}_{t,i,k}x^t\sum_{t=0}^{\overline{\overline{n}}_q}\overline{\overline{n}}_{t,k,l}x^t\sum_{t=0}^{\overline{\overline{l}}_q}\overline{\overline{l}}^*_{t,\mu,l}x^t}$$
$$\cdot\sum_{t=0}^{a_q}a_{t,\kappa,\lambda}x^t\sum_{t=0}^{a_q}a^*_{t,\kappa,\mu}x^t\sum_{t=0}^{a_q}a^*_{t,j,\lambda}x^t$$
$$= \sum_{\mu=1}^{n}\sum_{l=1}^{\min\{\mu,r\}}\sum_{k=1}^{\min\{i,r\}}\sum_{\lambda=1}^{n}\sum_{\kappa=1}^{m}\frac{\sum_{t=0}^{2\overline{l}_q+\overline{n}_q}\overline{p}_{t,i,k,l,\mu}x^t}{\sum_{t=0}^{2\overline{\overline{l}}_q+\overline{\overline{n}}_q}\overline{\overline{\beta}}_{t,i,k,l,\mu}x^t}\sum_{t=0}^{3a_q}\alpha_{t,j,\mu,\kappa,\lambda}x^t$$
$$= \sum_{\mu=1}^{n}\sum_{l=1}^{\min\{\mu,r\}}\sum_{k=1}^{\min\{i,r\}}\sum_{\lambda=1}^{n}\sum_{\kappa=1}^{m}\frac{\sum_{t=0}^{\overline{b}_q}\overline{\beta}_{t,i,j,k,l,\mu,\kappa,\lambda}x^t}{\sum_{t=0}^{\overline{\overline{b}}_q}\overline{\overline{\beta}}_{t,i,k,l,\mu}x^t}.$$

Accordingly, we have that $A_{ij}^{\dagger}(x) = \frac{\overline{\Gamma}_{ij}(x)}{\overline{\overline{\Gamma}}_i(x)}$, where is valid

$$\overline{\overline{\Gamma}}_i(x) = LCM\left\{\sum_{t=0}^{\overline{\overline{b}}_q}\overline{\overline{\beta}}_{t,i,k,l,\mu}x^t \,\middle|\, \mu=\overline{1,n}, k=\overline{1,\min\{i,r\}}, l=\overline{1,\min\{\mu,r\}}\right\}$$
$$= \sum_{t=0}^{\overline{\overline{\Gamma}}_q}\overline{\overline{\gamma}}_{t,i}x^t,$$
$$\overline{\Gamma}_{ij}(x) = \sum_{\mu=1}^{n}\sum_{l=1}^{\min\{\mu,r\}}\sum_{k=1}^{\min\{i,r\}}\left(\Gamma_{i,k,l,\mu}(x)\sum_{\lambda=1}^{n}\sum_{\kappa=1}^{m}\sum_{t=0}^{\overline{b}_q}\overline{\beta}_{t,i,j,k,l,\mu,\kappa,\lambda}x^t\right),$$

where each polynomial $\Gamma_{i,l,k,\mu}(x)$ is equal $\overline{\overline{\Gamma}}_i(x)/\left(\sum_{t=0}^{\overline{\overline{b}}_q}\overline{\overline{\beta}}_{t,i,k,l,\mu}x^t\right)=\sum_{t=0}^{\overline{\overline{\Gamma}}_q-\overline{\overline{b}}_q}\gamma_{t,i,k,l,\mu}x^t$. Hence,

$$\overline{\Gamma}_{ij}(x)=\sum_{\mu=1}^{n}\sum_{l=1}^{\min\{\mu,r\}}\sum_{k=1}^{\min\{i,r\}}\sum_{\lambda=1}^{n}\sum_{\kappa=1}^{m}\sum_{t=0}^{\overline{\overline{\Gamma}}_q-\overline{\overline{b}}_q+\overline{b}_q}\left(\sum_{t_1=0}^{t}\overline{\beta}_{t_1,i,j,k,l,\mu,\kappa,\lambda}\gamma_{t-t_1,i,k,l,\mu}\right)x^t,$$

which coincides with the form (2.3.26), which is a testament to complete. □

Obviously, it is possible to carry out a similar theorem watching the second assertion of Theorem 3.2.7. The obtained results it is possible to summarize the following algorithm.

Algorithm 3.3 Symbolic calculation of Moore–Penrose's inverse using LDL^* decomposition complete ranking

---

Input: Polynomial matrix $A(x)\in\mathbf{C}(x)_r^{m\times n}$ with the elements form $a_{ij}(x)=\sum_{t=0}^{a_q}a_{t,i,j}x^t$.

1: Generate LDL^* decomposition full-rank of matrix $(A^*A)^*(A^*A)$, where $L\in\mathbf{C}(x)^{n\times r}$ and $D\in\mathbf{C}(x)^{r\times r}$ matrices with the elements form (3.3.2), using the method given by equations from Theorem 3.3.1.

2: Transform rational matrix $M=L^*LDL^*L$ in form $M=\frac{1}{p(x)}M_1$, hence $p(x)$ polynomial M_1 is polynomial matrix

3: Compute the inverse matrix M_1 applying Algorithm 3.2 out of work [74]. Generate the inverse matrix $N=M^{-1}=p(x)M_1^{-1}$, and then reduced to the form:

$$n_{ij}(x)=\left(\sum_{k=0}^{\overline{n}_q}\overline{n}_{k,i,j}x^k\right)/\left(\sum_{k=0}^{\overline{\overline{n}}_q}\overline{\overline{n}}_{k,i,j}x^k\right).$$

4: For each $i = \overline{1,m},\ \mu = \overline{1,n},\ k = \overline{1,\min\{i,r\}},\ l = \overline{1,\min\{\mu,r\}}$ calculate the following ratios:

$$\overline{p}_{t,i,k,l,\mu} = \sum_{t_1=0}^{t}\sum_{t_2=0}^{t-t_1} \overline{l}_{t_1,i,k}\overline{n}_{t-t_1-t_2,k,l}\overline{l}^{*}_{t_2,\mu,l}, \tag{3.4.3}$$

$$0 \le t \le 2\overline{l}_q + \overline{n}_q \tag{3.4.4}$$

$$\overline{\overline{\beta}}_{t,i,k,l,\mu} = \sum_{t_1=0}^{t}\sum_{t_2=0}^{t-t_1} \overline{\overline{l}}_{t_1,i,k}\overline{\overline{n}}_{t-t_1-t_2,k,l}\overline{\overline{l}}^{*}_{t_2,\mu,l}, \tag{3.4.5}$$

$$0 \le t \le 2\overline{\overline{l}}_q + \overline{\overline{n}}_q. \tag{3.4.6}$$

5: For each $j = \overline{1,n},\ \mu = \overline{1,n},\ \kappa = \overline{1,m},\ \lambda = \overline{1,n}$ make the following evaluation:

$$\alpha_{t,j,\mu,\kappa,\lambda} = \sum_{t_1=0}^{t}\sum_{t_2=0}^{t-t_1} a_{t_1,\kappa,\lambda}a^{*}_{t-t_1-t_2,\kappa,\mu}a^{*}_{t_2,j,\lambda},\ 0 \le t \le 3a_q. \tag{3.4.7}$$

6: $\overline{b}_q = 2\overline{l}_q + \overline{n}_q + 3\overline{a}_q$, $\overline{\overline{b}}_q = 2\overline{\overline{l}}_q + \overline{\overline{n}}_q$ and for every $i = \overline{1,m},\ j = \overline{1,n},\ \mu = \overline{1,n},\ k = \overline{1,\min\{i,r\}},\ l = \overline{1,\min\{\mu,r\}},\ \kappa = \overline{1,m},\ \lambda = \overline{1,n}$ calculate

$$\overline{\beta}_{t,i,j,k,l,\mu,\kappa,\lambda} = \sum_{t_1=0}^{t} \overline{p}_{t_1,i,k,l,\mu}\alpha_{t-t_1,j,\mu,\kappa,\lambda}, \quad 0 \le t \le \overline{b}_q. \tag{3.4.8}$$

7: For $i = \overline{1,m}$evaluate polynomials in the denominator element $A^{\dagger}_{i,j}$ on the following way:

$$\overline{\overline{\Gamma}}_i(x) = LCM\left\{\sum_{t=0}^{\overline{\overline{b}}_q}\overline{\overline{\beta}}_{t,i,k,l,\mu}x^t \,\middle|\, \mu = \overline{1,n}, k = \overline{1,\min\{i,r\}}, l = \overline{1,\min\{\mu,r\}}\right\} \tag{3.4.9}$$

and mark with $\overline{\overline{\Gamma}}_i(x) = \sum_{t=0}^{\overline{\overline{\Gamma}}_q}\overline{\overline{\gamma}}_{t,i}x^t$.

8: For each $i = \overline{1,m},\ \mu = \overline{1,n},\ k = \overline{1,\min\{i,r\}},\ l = \overline{1,\min\{\mu,r\}}$ determine the following polynomial:

$$\overline{\overline{\Gamma}}_i(x) / \left(\sum_{t=0}^{\overline{\overline{b}}_q}\overline{\overline{\beta}}_{t,i,k,l,\mu}x^t\right),$$

and denote it as $\Gamma_{i,l,k,\mu}(x) = \sum_{t=0}^{\overline{\overline{\Gamma}}_q - \overline{\overline{b}}_q} \gamma_{t,i,k,l,\mu} x^t$.

9: For $i = \overline{1,m},\ j = \overline{1,n}$ calculate the numerator polynomial:

$$\overline{\Gamma}_{ij}(x) = \sum_{t=0}^{\overline{\overline{\Gamma}}_q - \overline{\overline{b}}_q + \overline{b}_q} \left(\sum_{\mu=1}^{n} \sum_{l=1}^{\min\{\mu,r\}} \sum_{k=1}^{\min\{i,r\}} \sum_{\lambda=1}^{n} \sum_{\kappa=1}^{m} \sum_{t_1=0}^{t} \overline{\beta}_{t_1,i,j,k,l,\mu,\kappa,\lambda} \gamma_{t-t_1,i,k,l,\mu} \right) x^t. \tag{3.4.10}$$

10: For $i = \overline{1,m},\ j = \overline{1,n}$ set (i,j)-th element Moore–Penrose inverse $A^{\dagger}$ na $\overline{\Gamma}_{ij}(x)/\overline{\overline{\Gamma}}_i(x)$.

LDL^* decomposition of the same complexity as the Cholesky decomposition. Note that LDL^* decomposition gives an additional diagonal matrix, or returns the result without square roots, suitable for further symbolic calculations. Also, the total number of non-zero elements is the same as for the Cholesky decomposition.

Note that even (i,j)-th element of matrix $(A^*A)^*(A^*A)$ be calculated as

$$\sum_{l=1}^{n} \sum_{k=1}^{m} \sum_{\kappa=1}^{m} a_{kl}^* a_{ki} a_{\kappa l}^* a_{\kappa j}.$$

Then, these polynomials input values Algorithm 3.3, used for the determination of LDL^* decomposition in Step 1 similar idea was used and in Step 3 above algorithm for determining the input values from Algorithm 3.2 [74].

3.4.1 Numerical Examples

Example 3.4.1. Let be given polynomial matrix S_3 from Ref. [88]. To calculate the matrix elements $A^{\dagger}$, LDL^* decomposition of the matrix $(S_3^* S_3)^* S_3^* S_3$ is determined by multiplying the direct elements of the matrix $L \in \mathbf{C}(x)^{3\times 2}$ and $D \in \mathbf{C}(x)^{2\times 2}$ based Algorithm 3.2. Thus,

$$L = \begin{bmatrix} 1 & 0 \\ \frac{5x+21x^2+27x^3+27x^4}{8+32x+57x^2+54x^3+27x^4} & 1 \\ 1 & 0 \end{bmatrix},$$

$$D = \begin{bmatrix} 8+32x+57x^2+54x^3+27x^4 & 0 \\ 0 & \frac{8}{8+32x+57x^2+54x^3+27x^4} \end{bmatrix}.$$

Following transformation matrix L^*LDL^*L in the manner described in the algorithm, applying Algorithm 3.2 from Ref. [74] we have:

$$N = (L^*LDL^*L)^{-1} = \begin{bmatrix} \frac{1}{32}\left(1-4x+12x^2+27x^4\right) & \frac{-85x-657x^2-2349x^3-5265x^4-7695x^5-8019x^6-5103x^7-2187x^8}{256+1024x+1824x^2+1728x^3+864x^4} \\ \frac{-85x-657x^2-2349x^3-5265x^4-7695x^5-8019x^6-5103x^7-2187x^8}{256+1024x+1824x^2+1728x^3+864x^4} & \frac{2048+24576x+142905x^2+532782x^3+1420335x^4+2858328x^5+4466826x^6+5484996x^7+5288166x^8+3936600x^9+2184813x^{10}+826686x^{11}+177147x^{12}}{2048+16384x+61952x^2+144384x^3+228384x^4+252288x^5+191808x^6+93312x^7+23328x^8} \end{bmatrix}$$

Evaluation coefficients from steps 4–6 are performed just as well as the evaluation of the polynomial least-common denominator needed in step 7. Note that the simplification of key importance in step 8, where calculated coefficients $\gamma_{t,i,k,l,\mu}$, $i=\overline{1,3}$, $k=\overline{1,2}$, $l=\overline{1,3}$, $\mu=\overline{1,3}$. Finally, generalized inverse

$$S_3^\dagger = \begin{bmatrix} \frac{1-x}{4} & \frac{x}{2} & \frac{1-x}{4} \\ \frac{x}{2} & -1-x & \frac{x}{2} \\ \frac{1-x}{4} & \frac{x}{2} & \frac{1-x}{4} \end{bmatrix}$$

is obtained after the simplification of each element is in the form of a fraction $\overline{\Gamma}_{ij}(x)/\overline{\overline{\Gamma}}_i(x)$, $i=\overline{1,3}$, $j=\overline{1,3}$,calculating the greatest common divisor of each pair of the numerator and denominator.

Example 3.4.2. Consider the following 4×3 polynomial matrix A_3 generated in Ref. [88]:

$$A_3 = \begin{bmatrix} 3+x & 2+x & 1+x \\ 2+x & 1+x & x \\ 1+x & x & -1+x \\ x & -1+x & -2+x \end{bmatrix}.$$

Given that the matrix A_3 is equal to 2, the LDL^* decomposition of the full-rank of the matrix $(A_3^*A_3)^*A_3^*A_3$ gains matrices $L \in \mathbf{C}(x)^{3\times 2}$ and $D \in$

$\mathbf{C}(x)^{2\times 2}$ with the following elements:

$$L=\begin{bmatrix} 1 & 0 \\ \frac{21+38x+37x^2+18x^3+6x^4}{33+60x+52x^2+24x^3+6x^4} & 1 \\ \frac{9+16x+22x^2+12x^3+6x^4}{33+60x+52x^2+24x^3+6x^4} & 2 \end{bmatrix},$$

$$D=\begin{bmatrix} 264+480x+416x^2+192x^3+48x^4 & 0 \\ 0 & \frac{300}{33+60x+52x^2+24x^3+6x^4} \end{bmatrix}.$$

Applying Algorithm 3.3 on matrices A_3, L and D, we get Moore–Penrose inverse matrices A_3:

$$A_3^{\dagger}=\begin{bmatrix} -\frac{3}{20}(-1+x) & \frac{1}{60}(8-3x) & \frac{1}{60}(7+3x) & \frac{1}{20}(2+3x) \\ \frac{1}{10} & \frac{1}{30} & -\frac{1}{30} & -\frac{1}{10} \\ \frac{1}{20}(1+3x) & \frac{1}{60}(-4+3x) & \frac{1}{60}(-11-3x) & -\frac{3}{20}(2+x) \end{bmatrix}.$$

3.4.2 Implementation Details

Based on formulation of Algorithm 3.3 it is possible to implement the following procedure for calculating Moore–Penrose's inverse polynomial matrix. Unique features the following argument is polynomial matrix A specified in a list. Note that the used functions `LDLDecomposition` to calculate LDL^* factorization matrix products $(A^*A)^2$.

```
LDLGInverse[A_List] := Module[{t, i, j, k, l, mu, m = Length[A], n = Length[A[[1]]],
   r = MatrixRank[A], L, D, N, GInv, f, p, beta1, beta2, alpha, kap, lam, r1},
  {L, D} = LDLDecomposition[Conjugate[Transpose[A]].A.Conjugate[Transpose[A]].A];
  L = ExpandDenominator[ExpandNumerator[L]];
  D = ExpandDenominator[ExpandNumerator[Together[D]]];
  N = Simplify[Inverse[Conjugate[Transpose[L]].L.D.Conjugate[Transpose[L]].L]];
  N = ExpandDenominator[ExpandNumerator[N]];
  p = Table[0,
    {2 * Max[Exponent[L, x]] + Max[Exponent[N, x]] + 1}, {m + 1}, {r + 1}, {r + 1}, {n + 1}];
  alpha = Table[0, {3 * Max[Exponent[A, x]] + 35}, {n + 1}, {n + 1}, {m + 1}, {n + 1}];
  beta1 = Table[0, {m}, {n}, {r}, {r}, {n}, {m}, {n}];
  beta2 = Table[0, {m}, {r}, {r}, {n}];
  GInv = Table[0, {m}, {n}];
  For[i = 1, i ≤ m, i++,
   For[k = 1, k ≤ Min[i, r], k++,
    For[l = 1, l ≤ Min[mu, r], l++,
     For[mu = 1, mu ≤ n, mu++,
      beta2[[i, k, l, mu]] = 0;
      For[t = 0, t ≤ Max[Exponent[L[[i, k]], x]] +
         Max[Exponent[N[[k, l]], x]] + Max[Exponent[L[[mu, l]], x]], t++,
       p[[t + 1, i, k, l, mu]] = ∑_{t1=0}^{t} ∑_{t2=0}^{t-t1} (Coefficient[Numerator[L[[i, k]]],
            x, t1] * Coefficient[Numerator[N[[k, l]]], x, t - t1 - t2] *
          Conjugate[Coefficient[Numerator[L[[mu, l]]], x, t2]]);
       beta2[[i, k, l, mu]] += x^t * ∑_{t1=0}^{t} ∑_{t2=0}^{t-t1} (Coefficient[Denominator[L[[i, k]]],
            x, t1] * Coefficient[Denominator[N[[k, l]]], x, t - t1 - t2] *
           Conjugate[Coefficient[Denominator[L[[mu, l]]], x, t2]]);
       ];];];];];
  For[j = 1, j ≤ n, j++,
   For[mu = 1, mu ≤ n, mu++,
    For[kap = 1, kap ≤ m, kap++,
     For[lam = 1, lam ≤ n, lam++,
      For[t = 0, t ≤ 3 * Max[Exponent[A, x]], t++,
       alpha[[t + 1, j, mu, kap, lam]] =
        ∑_{t1=0}^{t} ∑_{t2=0}^{t-t1} (Coefficient[A[[kap, lam]], x, t1] * Conjugate[Coefficient[A[[kap,
                mu]], x, t - t1 - t2]] * Conjugate[Coefficient[A[[j, lam]], x, t2]]);
       ];];];];];
```

```
For[i = 1, i ≤ m, i++,
  For[j = 1, j ≤ n, j++,
    For[k = 1, k ≤ Min[i, r], k++,
      For[l = 1, l ≤ Min[mu, r], l++,
        For[mu = 1, mu ≤ n, mu++,
          For[kap = 1, kap ≤ m, kap++,
            For[lam = 1, lam ≤ n, lam++,
              For[t = 0, t ≤ 2 * Max[Exponent[Numerator[L], x]] + Max[Exponent[
                    Numerator[N], x]] + 3 * Max[Exponent[Numerator[A], x]], t++,
                beta1[[i, j, k, l, mu, kap, lam]] += x^t * Sum[(p[[t1 + 1, i,
                      k, l, mu]] * alpha[[t - t1 + 1, j, mu, kap, lam]]), {t1, 0, t}];
              ];];];];];];];];
For[i = 1, i ≤ m, i++,
  For[j = 1, j ≤ n, j++,
    For[mu = 1, mu ≤ n, mu++,
      For[l = 1, l ≤ Min[mu, r], l++,
        For[k = 1, k ≤ Min[i, r], k++,
          br = beta2[[i, k, l, mu]] /. x -> 1;
          If[br ≠ 0,
            For[lam = 1, lam ≤ n, lam++,
              For[kap = 1, kap ≤ m, kap++,
                GInv[[i, j]] += Simplify[beta1[[i, j, k, l, mu, kap, lam]] / beta2[[i, k, l, mu]]];
              ];];
          ];];];];];
Return[Simplify[GInv]];];
```

3.5 LDL^* Decomposition of the Full-Rank of Rational Matrix

Some data is rational Hermitian matrix $A(s) \in \mathbb{C}[s]_r^{n\times n}$, with elements that represent the quotient of two polynomials undisclosed s:

$$a_{ij}(s) = \frac{\sum\limits_{k=0}^{\overline{a}_q} \overline{a}_{k,i,j} s^k}{\sum\limits_{k=0}^{\overline{\overline{a}}_q} \overline{\overline{a}}_{k,i,j} s^k}, \quad 1 \le i, j \le n. \tag{3.5.1}$$

Alternative form of Cholesky factorization, used in Ref. [56] means LDL^* decomposition of rational symmetric matrices $A(s)$ form

$$A(s) = L(s)D(s)L^*(s),$$

where $L(s)$ and $D(s)$ lower triangular and the diagonal matrix, respectively, and have the form given by

$$L(s) = \begin{bmatrix} 1 & 0 & \cdots & 0 \\ l_{21}(s) & 1 & \cdots & 0 \\ \vdots & \vdots & \ddots & \vdots \\ l_{n1}(s) & l_{n2}(s) & \cdots & 1 \end{bmatrix}, \qquad (3.5.2)$$

$$D(s) = \begin{bmatrix} d_1(s) & 0 & \cdots & 0 \\ 0 & d_2(s) & \cdots & 0 \\ \vdots & \vdots & \ddots & \vdots \\ 0 & 0 & \cdots & d_n(s) \end{bmatrix}. \qquad (3.5.3)$$

Using LDL^* factorization, avoids the calculation of elements with square roots. For example, work with expressions $\sqrt{\sum_{i=0}^{q} A_i s^i}$ symbolically, where A_i, $i = 0, \ldots, q$constant $n \times n$ matrix is very difficult. Because of this in Ref. [56] symbolic algorithm, for calculating the generalized inverse matrix of rational use of cleansing techniques is described. This method simplifies the problem symbolic calculations the rational generalized inverse matrix in procedural programming languages.

As previously observed, recurrence relations (3.3.3)–(3.3.5) need execute only $j = \overline{1, \text{rank}(A)}$. This method generates a representation of the full-rank of the matrix A, where L without zero column, a matrix D is not type contains zero and zero column. Accordingly, for a given rational matrix $A \in \mathbf{C}(s)_r^{m\times n} = \{X \in \mathbf{C}^{m\times n} \mid rank(X) = r\}$,consider decomposition $A = LDL^*$ full-rank, wherein $L \in \mathbf{C}(s)^{m\times r}$, $l_{ij} = 0$ for $i < j$, a $D \in \mathbf{C}(s)^{r\times r}$ is the rational diagonal matrix.

As matrices $L(s)$, $D(s)$ are rank r, they are in form

$$L(s) = \begin{bmatrix} 1 & 0 & \cdots & 0 \\ l_{2,1}(s) & 1 & \cdots & 0 \\ \vdots & \vdots & \ddots & \vdots \\ l_{r,1}(s) & l_{r,2}(s) & \cdots & 1 \\ l_{r+1,1}(s) & l_{r+1,2}(s) & \cdots & l_{r+1,r}(s) \\ \vdots & \vdots & \ddots & \vdots \\ l_{n,1}(s) & l_{n,2}(s) & \cdots & l_{n,r}(s) \end{bmatrix}, \tag{3.5.4}$$

$$D(s) = \begin{bmatrix} d_1(s) & 0 & \cdots & 0 \\ 0 & d_2(s) & \cdots & 0 \\ \vdots & \vdots & \ddots & \vdots \\ 0 & 0 & \cdots & d_r(s) \end{bmatrix}, \tag{3.5.5}$$

whereby the non-zero elements of rational matrix $L(s)$ and $D(s)$ form (3.3.2).

We will just relations (3.3.3)–(3.3.5) apply for directly calculating coefficients matrix elements in $L(s)$ and $D(s)$. As before, variable with one line will denote the coefficients of the polynomial in the numerator, and variables with two lines the coefficients of the polynomial in the denominator.

Equation (3.3.3) in the form of a polynomial can be written as:

$$\begin{aligned} f_{i,j}(s) &= \sum_{k=1}^{j-1} l_{ik}(s) l_{jk}^*(s) d_k(s) = \sum_{k=1}^{j-1} \frac{\sum_{t=0}^{\overline{l}_q} \overline{l}_{t,i,k} s^t}{\sum_{t=0}^{\overline{\overline{l}}_q} \overline{\overline{l}}_{t,i,k} s^t} \frac{\sum_{t=0}^{\overline{l}_q} \overline{l}^*_{t,j,k} s^t}{\sum_{t=0}^{\overline{\overline{l}}_q} \overline{\overline{l}}^*_{t,j,k} s^t} \frac{\sum_{t=0}^{\overline{d}_q} \overline{d}_{t,k} s^t}{\sum_{t=0}^{\overline{\overline{d}}_q} \overline{\overline{d}}_{t,k} s^t} \\ &= \sum_{k=1}^{j-1} \frac{\sum_{r_1=0}^{2\overline{l}_q+\overline{d}_q} \left(\sum_{r_2=0}^{r_1} \sum_{r_3=0}^{r_1-r_2} \overline{l}_{r_3,i,k} \overline{l}^*_{r_1-r_2-r_3,j,k} \overline{d}_{r_2,k} \right) s^{r_1}}{\sum_{r_1=0}^{2\overline{\overline{l}}_q+\overline{\overline{d}}_q} \left(\sum_{r_2=0}^{r_1} \sum_{r_3=0}^{r_1-r_2} \overline{\overline{l}}_{r_3,i,k} \overline{\overline{l}}^*_{r_1-r_2-r_3,j,k} \overline{\overline{d}}_{r_2,k} \right) s^{r_1}}. \end{aligned}$$

Let us introduce the following codes:

$$\begin{aligned}
\overline{p}_{r_1,i,j,k} &= \sum_{r_2=0}^{r_1}\sum_{r_3=0}^{r_1-r_2} \overline{l}_{r_3,i,k}\overline{l}^{*}_{r_1-r_2-r_3,j,k}\overline{d}_{r_2,k},\ 0\le r_1\le 2\overline{l}_q+\overline{d}_q,\\
\overline{\overline{p}}_{r_1,i,j,k} &= \sum_{r_2=0}^{r_1}\sum_{r_3=0}^{r_1-r_2} \overline{\overline{l}}_{r_3,i,k}\overline{\overline{l}}^{*}_{r_1-r_2-r_3,j,k}\overline{\overline{d}}_{r_2,k},\ 0\le r_1\le 2\overline{\overline{l}}_q+\overline{\overline{d}}_q.
\end{aligned}$$

Then satisfied

$$\begin{aligned}
f_{i,j}(s) &= \sum_{k=1}^{j-1} \frac{\sum_{r_1=0}^{2\overline{l}_q+\overline{d}_q} \overline{p}_{r_1,i,j,k}s^{r_1}}{\sum_{r_1=0}^{2\overline{\overline{l}}_q+\overline{\overline{d}}_q} \overline{\overline{p}}_{r_1,i,j,k}s^{r_1}}\\
&= \frac{\sum_{r_1=0}^{2\overline{l}_q+\overline{d}_q} \overline{p}_{r_1,i,j,1}s^{r_1}}{\sum_{r_1=0}^{2\overline{\overline{l}}_q+\overline{\overline{d}}_q} \overline{\overline{p}}_{r_1,i,j,1}s^{r_1}} + \frac{\sum_{r_1=0}^{2\overline{l}_q+\overline{d}_q} \overline{p}_{r_1,i,j,2}s^{r_1}}{\sum_{r_1=0}^{2\overline{\overline{l}}_q+\overline{\overline{d}}_q} \overline{\overline{p}}_{r_1,i,j,2}s^{r_1}}\\
&\quad + \ldots + \frac{\sum_{r_1=0}^{2\overline{l}_q+\overline{d}_q} \overline{p}_{r_1,i,j,j-1}s^{r_1}}{\sum_{r_1=0}^{2\overline{\overline{l}}_q+\overline{\overline{d}}_q} \overline{\overline{p}}_{r_1,i,j,j-1}s^{r_1}}.
\end{aligned}$$

Now determine the lowest common denominator (GSC) in the denominator polynomial and mark it as follows:

$$\begin{aligned}
LCM &\left(\sum_{k=0}^{2\overline{\overline{l}}_q+\overline{\overline{d}}_q} \overline{\overline{p}}_{k,i,j,1}s^k, \sum_{k=0}^{2\overline{\overline{l}}_q+\overline{\overline{d}}_q} \overline{\overline{p}}_{k,i,j,2}s^k, \ldots, \sum_{k=0}^{2\overline{\overline{l}}_q+\overline{\overline{d}}_q} \overline{\overline{p}}_{k,i,j,j-1}s^k \right)\\
&= \sum_{k=0}^{\overline{\overline{f}}_q} \overline{\overline{f}}_{k,i,j}s^k.
\end{aligned}$$

Also let us introduce labels

$$q_{i,j,t}(s) = \frac{\sum_{k=0}^{\overline{\overline{f}}_q} \overline{\overline{f}}_{k,i,j} s^k}{\sum_{k=0}^{2\overline{l}_q+\overline{\overline{d}}_q} \overline{\overline{p}}_{k,i,j,t} s^k} = \sum_{k=0}^{\overline{\overline{f}}_q-2\overline{l}_q-\overline{\overline{d}}_q} q_{k,i,j,t} s^k,\ 1 \le t \le j-1,\ i > j.$$

Let now be $\overline{f}_q = 2\overline{l}_q + \overline{d}_q + \overline{\overline{f}}_q - 2\overline{l}_q - \overline{\overline{d}}_q$. Continuing with calculation elements $f_{i,j}(s)$ we have

$$\begin{aligned} f_{i,j}(s) &= \frac{\sum_{k=1}^{j-1} \sum_{k_1=0}^{\overline{f}_q} \left(\sum_{k_2=0}^{k_1} \overline{p}_{k_1-k_2,i,j,k} q_{k_2,i,j,k} \right) s^{k_1}}{\sum_{k=0}^{\overline{\overline{f}}_q} \overline{\overline{f}}_{k,i,j} s^k} \\ &= \frac{\sum_{k_1=0}^{\overline{f}_q} \sum_{k_2=0}^{k_1} \sum_{k_3=0}^{j-1} \overline{p}_{k_1-k_2,i,j,k_3} q_{k_2,i,j,k_3} s^{k_1}}{\sum_{k=0}^{\overline{\overline{f}}_q} \overline{\overline{f}}_{k,i,j} s^k} = \frac{\sum_{k=0}^{\overline{f}_q} \overline{f}_{k,i,j} s^k}{\sum_{k=0}^{\overline{\overline{f}}_q} \overline{\overline{f}}_{k,i,j} s^k}, \end{aligned}$$

where $\overline{f}_{k,i,j} = \sum_{k_2=0}^{k} \sum_{k_3=0}^{j-1} \overline{p}_{k-k_2,i,j,k_3} q_{k_2,i,j,k_3},\ 0 \le k \le \overline{f}_q$.

Notice, now that for every $j=\overline{1,r}$, based on equalities (3.3.4), proceed the next few evaluations

$$\begin{aligned}
d_j(s) &= \frac{\sum_{k=0}^{\overline{a}_q} \overline{a}_{k,j,j}s^k}{\sum_{k=0}^{\overline{\overline{a}}_q} \overline{\overline{a}}_{k,j,j}s^k} - \frac{\sum_{k=0}^{\overline{f}_q} \overline{f}_{k,j,j}s^k}{\sum_{k=0}^{\overline{\overline{f}}_q} \overline{\overline{f}}_{k,j,j}s^k} \\
&= \frac{\sum_{k=0}^{\overline{a}_q} \overline{a}_{k,j,j}s^k \sum_{k=0}^{\overline{\overline{f}}_q} \overline{\overline{f}}_{k,j,j}s^k - \sum_{k=0}^{\overline{\overline{a}}_q} \overline{\overline{a}}_{k,j,j}s^k \sum_{k=0}^{\overline{f}_q} \overline{f}_{k,j,j}s^k}{\sum_{k=0}^{\overline{\overline{a}}_q} \overline{\overline{a}}_{k,j,j}s^k \sum_{k=0}^{\overline{\overline{f}}_q} \overline{\overline{f}}_{k,j,j}s^k} \\
&= \frac{\sum_{k_1=0}^{\overline{a}_q+\overline{\overline{f}}_q} \left(\sum_{k_2=0}^{k_1} \overline{a}_{k_1-k_2,j,j}\overline{\overline{f}}_{k_2,j,j} \right) s^{k_1} - \sum_{k_1=0}^{\overline{\overline{a}}_q+\overline{f}_q} \left(\sum_{k_2=0}^{k_1} \overline{\overline{a}}_{k_1-k_2,j,j}\overline{f}_{k_2,j,j} \right) s^{k_1}}{\sum_{k=0}^{\overline{\overline{a}}_q} \overline{\overline{a}}_{k,j,j}s^k \sum_{k=0}^{\overline{\overline{f}}_q} \overline{\overline{f}}_{k,j,j}s^k} \\
&= \frac{\sum_{k_1=0}^{\max(\overline{a}_q+\overline{\overline{f}}_q,\overline{\overline{a}}_q+\overline{f}_q)} \left(\sum_{k_2=0}^{k_1} (\overline{a}_{k_1-k_2,j,j}\overline{\overline{f}}_{k_2,j,j} - \overline{\overline{a}}_{k_1-k_2,j,j}\overline{f}_{k_2,j,j}) \right) s^{k_1}}{\sum_{k_1=0}^{\overline{\overline{a}}_q+\overline{\overline{f}}_q} \left(\sum_{k_2=0}^{k_1} \overline{\overline{a}}_{k_1-k_2,j,j}\overline{\overline{f}}_{k_2,j,j} \right) s^{k_1}} \\
&= \frac{\sum_{k=0}^{\overline{d}_q} \overline{d}_{k,j}s^k}{\sum_{k=0}^{\overline{\overline{d}}_q} \overline{\overline{d}}_{k,j}s^k},
\end{aligned}$$

wherein

$$\begin{aligned}
\overline{d}_q &= \max(\overline{a}_q+\overline{\overline{f}}_q,\overline{\overline{a}}_q+\overline{f}_q), \quad \overline{\overline{d}}_q = \overline{\overline{a}}_q+\overline{\overline{f}}_q, \\
\overline{d}_{k,j} &= \sum_{i=0}^{k} (\overline{a}_{k-i,j,j}\overline{\overline{f}}_{i,j,j} - \overline{\overline{a}}_{k-i,j,j}\overline{f}_{i,j,j}), \; 0 \le k \le \overline{d}_q, \\
\overline{\overline{d}}_{k,j} &= \sum_{i=0}^{k} \overline{\overline{a}}_{k-i,j,j}\overline{\overline{f}}_{i,j,j}, \; 0 \le k \le \overline{\overline{d}}_q.
\end{aligned}$$

Finally, on the basis of equality (3.3.5), it is satisfied the following

$$
\begin{aligned}
l_{i,j}(s) &= \frac{\sum\limits_{k=0}^{\overline{\overline{d}}_q} \overline{\overline{d}}_{k,j}s^k}{\sum\limits_{k=0}^{\overline{d}_q} \overline{d}_{k,j}s^k} \left(\frac{\sum\limits_{k=0}^{\overline{a}_q} \overline{a}_{k,i,j}s^k}{\sum\limits_{k=0}^{\overline{\overline{a}}_q} \overline{\overline{a}}_{k,i,j}s^k} - \frac{\sum\limits_{k=0}^{\overline{f}_q} \overline{f}_{k,i,j}s^k}{\sum\limits_{k=0}^{\overline{\overline{f}}_q} \overline{\overline{f}}_{k,i,j}s^k} \right) \\
&= \frac{\sum\limits_{k=0}^{\overline{\overline{d}}_q} \overline{\overline{d}}_{k,j}s^k}{\sum\limits_{k=0}^{\overline{d}_q} \overline{d}_{k,j}s^k} \cdot \frac{\sum\limits_{k_1=0}^{\max(\overline{a}_q+\overline{\overline{f}}_q, \overline{\overline{a}}_q+\overline{f}_q)} \left(\sum\limits_{k_2=0}^{k_1} (\overline{a}_{k_1-k_2,i,j}\overline{\overline{f}}_{k_2,i,j} - \overline{\overline{a}}_{k_1-k_2,i,j}\overline{f}_{k_2,i,j}) \right) s^{k_1}}{\sum\limits_{k_1=0}^{\overline{\overline{a}}_q+\overline{\overline{f}}_q} \left(\sum\limits_{k_2=0}^{k_1} \overline{\overline{a}}_{k_1-k_2,i,j}\overline{\overline{f}}_{k_2,i,j} \right) s^{k_1}} \\
&= \frac{\sum\limits_{k=0}^{\overline{\overline{d}}_q+\max(\overline{a}_q+\overline{\overline{f}}_q, \overline{\overline{a}}_q+\overline{f}_q)} \left(\sum\limits_{k_1=0}^{k} \overline{\overline{d}}_{k_1,j} \left(\sum\limits_{k_2=0}^{k-k_1} (\overline{a}_{k-k_1-k_2,i,j}\overline{\overline{f}}_{k_2,i,j} - \overline{\overline{a}}_{k-k_1-k_2,i,j}\overline{f}_{k_2,i,j}) \right) \right) s^k}{\sum\limits_{k=0}^{\overline{d}_q+\overline{\overline{a}}_q+\overline{\overline{f}}_q} \left(\sum\limits_{k_1=0}^{k} \overline{d}_{k_1,j} \left(\sum\limits_{k_2=0}^{k-k_1} \overline{\overline{a}}_{k-k_1-k_2,i,j}\overline{\overline{f}}_{k_2,i,j} \right) \right) s^k} \\
&= \frac{\sum\limits_{k=0}^{\overline{l}_q} \overline{l}_{k,i,j}s^k}{\sum\limits_{k=0}^{\overline{\overline{l}}_q} \overline{\overline{l}}_{k,i,j}s^k},
\end{aligned}
$$

where

$$
\begin{aligned}
\overline{l}_q &= \overline{\overline{d}}_q + \max(\overline{a}_q+\overline{\overline{f}}_q, \overline{\overline{a}}_q+\overline{f}_q), \quad \overline{\overline{l}}_q = \overline{d}_q + \overline{\overline{a}}_q + \overline{\overline{f}}_q \\
\overline{l}_{k,i,j} &= \sum_{k_1=0}^{k} \overline{\overline{d}}_{k_1,j} \left(\sum_{k_2=0}^{k-k_1} (\overline{a}_{k-k_1-k_2,i,j}\overline{\overline{f}}_{k_2,i,j} - \overline{\overline{a}}_{k-k_1-k_2,i,j}\overline{f}_{k_2,i,j}) \right), \\
& \quad 0 \le k \le \overline{l}_q, \\
\overline{\overline{l}}_{k,i,j} &= \sum_{k_1=0}^{k} \overline{d}_{k_1,j} \left(\sum_{k_2=0}^{k-k_1} \overline{\overline{a}}_{k-k_1-k_2,i,j}\overline{\overline{f}}_{k_2,i,j} \right), \quad 0 \le k \le \overline{\overline{l}}_q.
\end{aligned}
$$

Next algorithm to calculate LDL^* decomposition of $A(s) = \{a_{ij}(s)\}_{i,j=1}^{n}$ which is Hermitian polynomial matrix, is introduced in the Ref. [56].

Algorithm 3.4 LDL^*factorization full-rank of rational Hermitian matrix

---

Input: Rational Hermitian matrix $A(s) \in \mathbf{C}[s]_r^{n\times n}$.

1: Initialization: $d_1(s) := a_{11}(s)$ i $l_{i1}(s) := \frac{a_{i1}(s)}{a_{11}(s)}$ for $i = \overline{2,n}$. Set $L_{ii}(s) :=$ 1, for every $i = \overline{1,n}$. For $j = \overline{2,r}$ execute steps 2, 3, 4.

2: Calculate the coefficients of the polynomial $f_{ij}(s)$ through the following four steps, i.e., for $i = \overline{j,n}$ execute Steps 2.1–2.4.

2.1: For every $k = \overline{1, j-1}$, make the following calculations equalization:

$$\overline{p}_{r_1,i,j,k} = \sum_{r_2=0}^{r_1}\sum_{r_3=0}^{r_1-r_2} \overline{l}_{r_3,i,k}\overline{l}^*_{r_1-r_2-r_3,j,k}\overline{d}_{r_2,k}, \quad 0 \le r_1 \le 2\overline{l}_q + \overline{d}_q,$$

$$\overline{\overline{p}}_{r_1,i,j,k} = \sum_{r_2=0}^{r_1}\sum_{r_3=0}^{r_1-r_2} \overline{\overline{l}}_{r_3,i,k}\overline{\overline{l}}^*_{r_1-r_2-r_3,j,k}\overline{\overline{d}}_{r_2,k}, \quad 0 \le r_1 \le 2\overline{\overline{l}}_q + \overline{\overline{d}}_q.$$

2.2: Determine the lowest common denominator polynomial denominators:

$$\sum_{k=0}^{\overline{\overline{f}}_q} \overline{\overline{f}}_{k,i,j}s^k = LCM\left(\sum_{k=0}^{2\overline{\overline{l}}_q+\overline{\overline{d}}_q} \overline{\overline{p}}_{k,i,j,1}s^k, \ldots, \sum_{k=0}^{2\overline{\overline{l}}_q+\overline{\overline{d}}_q} \overline{\overline{p}}_{k,i,j,j-1}s^k\right).$$

2.3: For every $t = \overline{1, j-1}$ i $i = \overline{j+1, n}$ assign the resulting polynomial $\sum_{k=0}^{\overline{\overline{f}}_q} \overline{\overline{f}}_{k,i,j}s^k$ with the following polynomial $\sum_{k=0}^{2\overline{\overline{l}}_q+\overline{\overline{d}}_q} \overline{\overline{p}}_{k,i,j,t}s^k$, and label quotient as

$$q_{i,j,t}(s) = \sum_{k=0}^{\overline{\overline{f}}_q - 2\overline{\overline{l}}_q - \overline{\overline{d}}_q} q_{k,i,j,t}s^k.$$

2.4: Introduce substitution $\overline{f}_q = 2\overline{l}_q + \overline{d}_q + \overline{\overline{f}}_q - 2\overline{\overline{l}}_q - \overline{\overline{d}}_q$, and calculate

$$\overline{f}_{k,i,j} = \sum_{k_2=0}^{k} \sum_{k_3=0}^{j-1} \overline{p}_{k-k_2,i,j,k_3} q_{k_2,i,j,k_3},\ 0 \leq k \leq \overline{f}_q.$$

3: Introduce the following substitutions: $\overline{d}_q = \max(\overline{a}_q + \overline{\overline{f}}_q, \overline{\overline{a}}_q + \overline{f}_q)$ and $\overline{\overline{d}}_q = \overline{\overline{a}}_q + \overline{\overline{f}}_q$, and calculate

$$\begin{aligned} \overline{d}_{k,j} &= \sum_{i=0}^{k} (\overline{a}_{k-i,j,j} \overline{\overline{f}}_{i,j,j} - \overline{\overline{a}}_{k-i,j,j} \overline{f}_{i,j,j}),\ 0 \leq k \leq \overline{d}_q, \\ \overline{\overline{d}}_{k,j} &= \sum_{i=0}^{k} \overline{\overline{a}}_{k-i,j,j} \overline{\overline{f}}_{i,j,j},\ 0 \leq k \leq \overline{\overline{d}}_q. \end{aligned}$$

4: Introduce substitutions $\overline{l}_q = \overline{\overline{d}}_q + \max(\overline{a}_q + \overline{\overline{f}}_q, \overline{\overline{a}}_q + \overline{f}_q)$ and $\overline{\overline{l}}_q = \overline{d}_q + \overline{\overline{a}}_q + \overline{\overline{f}}_q$, a for each $i = \overline{j+1,n}$ calculate the following:

$$\begin{aligned} \overline{l}_{k,i,j} &= \sum_{k_1=0}^{k} \overline{\overline{d}}_{k_1,j} \left(\sum_{k_2=0}^{k-k_1} (\overline{a}_{k-k_1-k_2,i,j} \overline{\overline{f}}_{k_2,i,j} - \overline{\overline{a}}_{k-k_1-k_2,i,j} \overline{f}_{k_2,i,j}) \right),\ 0 \leq k \leq \overline{l}_q, \\ \overline{\overline{l}}_{k,i,j} &= \sum_{k_1=0}^{k} \overline{d}_{k_1,j} \left(\sum_{k_2=0}^{k-k_1} \overline{\overline{a}}_{k-k_1-k_2,i,j} \overline{\overline{f}}_{k_2,i,j} \right),\ 0 \leq k \leq \overline{\overline{l}}_q. \end{aligned}$$

5: **Rezultat:**

$$d_j(s) = \frac{\sum_{k=0}^{\overline{d}_q} \overline{d}_{k,j} s^k}{\sum_{k=0}^{\overline{\overline{d}}_q} \overline{\overline{d}}_{k,j} s^k},\quad l_{i,j}(s) = \frac{\sum_{k=0}^{\overline{l}_q} \overline{l}_{k,i,j} s^k}{\sum_{k=0}^{\overline{\overline{l}}_q} \overline{\overline{l}}_{k,i,j} s^k},\quad 1 \leq i \leq n,\ 1 \leq j \leq r.$$

Implementation of Algorithm 3.4 is very similar to the implementation of the Algorithm 3.2. Weather calculations when applying Algorithm 3.4is not opposed to its polynomial version of the conference polynomial test matrix. Due to the complexity of the algorithm is LDL^* decomposition in the worst case $O(n^3)$ complexity Algorithm 3.4 will be approximately $O(n^3m^2)$, where m is the maximum exponent of the polynomial algorithm obtained.

Notice that in some test examples, polynomial coefficients can expand greatly during run-time, in spite to the performed simplification. One possible solution to this problem in procedural programming languages, is to implement the large number operations. The second possibility to handle this is to implement entries being the quotients of two polynomial, and to do the simplification after each step.

3.6 Calculating the Moore–Penrose's Inverse Rational Matrix

The motivation is to use no decomposition algorithm to obtain root for the calculation of Moore–Penrose's inverse rational matrix. Such methods presented in Ref. [56] is very applicable in pogramming procedural languages, as well as symbolic programming packages, such as `MATHEMATICA`. Next, the result of Theorem 3.2.7 over there in case of rational matrix is introduced in Ref. [56].

Theorem 3.6.1. [56] *Consider a rational matrix* $A \in \mathbf{C}(x)_r^{m\times n}$ *i* LDL^* *full-rank decomposition of matrix* $(A^*A)^*(A^*A)$*, whereby the an arbitrary element of the inverse matrix* $N = (L^*LDL^*L)^{-1} \in \mathbf{C}(x)^{r\times r}$ *denoted as*

$$n_{ij}(x) = \left(\sum_{t=0}^{\overline{n}_q} \overline{n}_{t,i,j}x^t\right) / \left(\sum_{t=0}^{\overline{\overline{n}}_q} \overline{\overline{n}}_{t,i,j}x^t\right). \tag{3.6.1}$$

Some have been introduced for the following tags $\mu = \overline{1,n}, k = \overline{1,\min\{i,r\}}, l = \overline{1,\min\{\mu,r\}}, \kappa = \overline{1,m}, \lambda = \overline{1,n}$*:*

$$\overline{\beta}_{t,i,j,k,l,\mu,\kappa,\lambda} = \sum_{t_1=0}^{t} \overline{p}_{t_1,i,k,l,\mu}\overline{\alpha}_{t-t_1,j,\mu,\kappa,\lambda}, \tag{3.6.2}$$

$$0 \le t \le \overline{b}_q = 2\overline{l}_q + \overline{n}_q + 3\overline{a}_q, \tag{3.6.3}$$

$$\overline{\overline{\beta}}_{t,i,j,k,l,\mu,\kappa,\lambda} = \sum_{t_1=0}^{t} \overline{\overline{p}}_{t_1,i,k,l,\mu}\overline{\overline{\alpha}}_{t-t_1,j,\mu,\kappa,\lambda}, \tag{3.6.4}$$

$$0 \le t \le \overline{\overline{b}}_q = 2\overline{\overline{l}}_q + \overline{\overline{n}}_q + 3\overline{\overline{a}}_q, \tag{3.6.5}$$

$$\overline{p}_{t,i,k,l,\mu} = \sum_{t_1=0}^{t}\sum_{t_2=0}^{t-t_1} \overline{l}_{t_1,i,k}\overline{n}_{t-t_1-t_2,k,l}\overline{l}^{*}_{t_2,\mu,l}, \tag{3.6.6}$$

$$0 \le t \le 2\overline{l}_q + \overline{n}_q \tag{3.6.7}$$

$$\overline{\overline{p}}_{t,i,k,l,\mu} = \sum_{t_1=0}^{t}\sum_{t_2=0}^{t-t_1} \overline{\overline{l}}_{t_1,i,k}\overline{\overline{n}}_{t-t_1-t_2,k,l}\overline{\overline{l}}^{*}_{t_2,\mu,l}, \tag{3.6.8}$$

$$0 \le t \le= 2\overline{\overline{l}}_q + \overline{\overline{n}}_q, \tag{3.6.9}$$

$$\overline{\alpha}_{t,j,\mu,\kappa,\lambda} = \sum_{t_1=0}^{t}\sum_{t_2=0}^{t-t_1} \overline{a}_{t_1,\kappa,\lambda}\overline{a}^{*}_{t-t_1-t_2,\kappa,\mu}\overline{a}^{*}_{t_2,j,\lambda}, \tag{3.6.10}$$

$$0 \le t \le 3\overline{a}_q, \tag{3.6.11}$$

$$\alpha_{t,j,\mu,\kappa,\lambda} = \sum_{t_1=0}^{t}\sum_{t_2=0}^{t-t_1} \overline{\overline{a}}_{t_1,\kappa,\lambda}\overline{\overline{a}}^{*}_{t-t_1-t_2,\kappa,\mu}\overline{\overline{a}}^{*}_{t_2,j,\lambda}, \tag{3.6.12}$$

$$0 \le t \le 3\overline{\overline{a}}_q. \tag{3.6.13}$$

Then the an arbitrary (i,j)*-th element of Moore–Penrose's inverse of matrix A is given with*

$$A^{\dagger}_{ij}(x) = \frac{\Phi_{ij}(x)}{\Psi_{ij}(x)},$$

wherein

$$\Phi_{ij}(x)=\sum_{\mu=1}^{n}\sum_{l=1}^{\min\{\mu,r\}}\sum_{k=1}^{\min\{i,r\}}\sum_{\lambda=1}^{n}\sum_{\kappa=1}^{m}\sum_{t=0}^{\Psi_q-\overline{\overline{b}}_q+\overline{b}_q}\left(\sum_{t_1=0}^{t}\overline{\beta}_{t_1,i,j,k,l,\mu,\kappa,\lambda}\gamma_{t-t_1,i,j,k,l,\mu,\kappa,\lambda}\right)x^t,$$

$$\Psi_{ij}(x)=LCM\{\sum_{t=0}^{\overline{\overline{b}}_q}\overline{\overline{\beta}}_{t,i,j,k,l,\mu,\kappa,\lambda}x^t\,|$$

$$k=\overline{1,\min\{i,r\}},l=\overline{1,\min\{\mu,r\}},\mu=\overline{1,n},\kappa=\overline{1,m},\lambda=\overline{1,n}\}, \tag{3.6.14}$$

while $\gamma_{t,i,j,k,l,\mu,\kappa,\lambda}$, $0\le t\le\Psi_q-\overline{\overline{b}}_q$, *coefficients of polynomial*

$$\Gamma_{i,j,k,l,\mu,\kappa,\lambda}(x)=\frac{\Psi_{ij}(x)}{\sum_{t=0}^{\overline{\overline{b}}_q}\overline{\overline{\beta}}_{t,i,j,k,l,\mu,\kappa,\lambda}x^t},$$

and Ψ_q *is the maximum exponent of the polynomial* $\Psi_{ij}(x)$.

Proof. Suppose, without impairing generality, that the rational elements of the matrix A given in the form (3.5.1). How LDL^* represent the full-rank of the matrix decomposition $(A^*A)^*(A^*A)\in\mathbf{C}(x)_r^{n\times n}$, follows that $L\in\mathbf{C}(x)^{n\times r}$ i $D\in\mathbf{C}(x)^{r\times r}$ matrix elements given in the form of (3.3.2).

Based on the statement of the theorem 3.2.7, an arbitrary (i,j)-th element of matrix $A^\dagger$ can be calculated as:

$$\begin{aligned}A^\dagger_{ij} &= \sum_{\mu=1}^{n}(LNL^*)_{i\mu}((A^*A)^*A^*)_{\mu j}\\ &= \sum_{\mu=1}^{n}\left(\sum_{l=1}^{\min\{\mu,r\}}\sum_{k=1}^{\min\{i,r\}}l_{ik}n_{kl}l^*_{\mu l}\right)\left(\sum_{\lambda=1}^{n}\sum_{\kappa=1}^{m}a_{\kappa\lambda}a^*_{\kappa\mu}a^*_{j\lambda}\right)\\ &= \sum_{\mu=1}^{n}\sum_{l=1}^{\min\{\mu,r\}}\sum_{k=1}^{\min\{i,r\}}\sum_{\lambda=1}^{n}\sum_{\kappa=1}^{m}l_{ik}n_{kl}l^*_{\mu l}a_{\kappa\lambda}a^*_{\kappa\mu}a^*_{j\lambda}.\end{aligned}$$

By taking rational forms of matrix elements, (i,j)-th element of Moore–Penrose inverse of A can be determined.

$$A^{\dagger}_{ij}(x) = \sum_{\mu=1}^{n}\sum_{l=1}^{\min\{\mu,r\}}\sum_{k=1}^{\min\{i,r\}}\sum_{\lambda=1}^{n}\sum_{\kappa=1}^{m} \frac{\sum_{t=0}^{\overline{l}_q}\overline{l}_{t,i,k}x^t}{\sum_{t=0}^{\overline{\overline{l}}_q}\overline{\overline{l}}_{t,i,k}x^t}\cdot\frac{\sum_{t=0}^{\overline{n}_q}\overline{n}_{t,k,l}x^t}{\sum_{t=0}^{\overline{\overline{n}}_q}\overline{\overline{n}}_{t,k,l}x^t}\cdot$$

$$\frac{\sum_{t=0}^{\overline{l}_q}\overline{l}^*_{t,\mu,l}x^t}{\sum_{t=0}^{\overline{\overline{l}}_q}\overline{\overline{l}}^*_{t,\mu,l}x^t}\cdot\frac{\sum_{t=0}^{\overline{a}_q}\overline{a}_{t,\kappa,\lambda}x^t}{\sum_{t=0}^{\overline{\overline{a}}_q}\overline{\overline{a}}_{t,\kappa,\lambda}x^t}\cdot\frac{\sum_{t=0}^{\overline{a}_q}\overline{a}^*_{t,\kappa,\mu}x^t}{\sum_{t=0}^{\overline{\overline{a}}_q}\overline{\overline{a}}^*_{t,\kappa,\mu}x^t}\cdot\frac{\sum_{t=0}^{\overline{a}_q}\overline{a}^*_{t,j,\lambda}x^t}{\sum_{t=0}^{\overline{\overline{a}}_q}\overline{\overline{a}}^*_{t,j,\lambda}x^t}$$

$$= \sum_{\mu=1}^{n}\sum_{l=1}^{\min\{\mu,r\}}\sum_{k=1}^{\min\{i,r\}}\sum_{\lambda=1}^{n}\sum_{\kappa=1}^{m} \frac{\sum_{t=0}^{2\overline{l}_q+\overline{n}_q}\overline{p}_{t,i,k,l,\mu}x^t\ \sum_{t=0}^{3\overline{a}_q}\overline{\alpha}_{t,j,\mu,\kappa,\lambda}x^t}{\sum_{t=0}^{2\overline{\overline{l}}_q+\overline{\overline{n}}_q}\overline{\overline{p}}_{t,i,k,l,\mu}x^t\ \sum_{t=0}^{3\overline{\overline{a}}_q}\overline{\overline{\alpha}}_{t,j,\mu,\kappa,\lambda}x^t}$$

$$= \sum_{\mu=1}^{n}\sum_{l=1}^{\min\{\mu,r\}}\sum_{k=1}^{\min\{i,r\}}\sum_{\lambda=1}^{n}\sum_{\kappa=1}^{m} \frac{\sum_{t=0}^{\overline{b}_q}\overline{\beta}_{t,i,j,k,l,\mu,\kappa,\lambda}x^t}{\sum_{t=0}^{\overline{\overline{b}}_q}\overline{\overline{\beta}}_{t,i,j,k,l,\mu,\kappa,\lambda}x^t}.$$

Therefore, $A^{\dagger}_{ij}(x) = \frac{\Phi_{ij}(x)}{\Psi_{ij}(x)}$ is valid, where

$$\Psi_{ij}(x) = LCM\{\sum_{t=0}^{\overline{\overline{b}}_q}\overline{\overline{\beta}}_{t,i,j,k,l,\mu,\kappa,\lambda}x^t\},$$

$where, \mu=\overline{1,n}, k=\overline{1,\min\{i,r\}}, l=\overline{1,\min\{\mu,r\}}, \kappa=\overline{1,m}, \lambda=\overline{1,n}.$

$$\Phi_{ij}(x) = \sum_{\mu=1}^{n}\sum_{l=1}^{\min\{\mu,r\}}\sum_{k=1}^{\min\{i,r\}}\left(\Gamma_{i,j,k,l,\mu,\kappa,\lambda}(x)\sum_{\lambda=1}^{n}\sum_{\kappa=1}^{m}\sum_{t=0}^{\overline{b}_q}\overline{\beta}_{t,i,j,k,l,\mu,\kappa,\lambda}x^t\right),$$

and polynomial $\Gamma_{i,j,k,l,\mu,\kappa,\lambda}(x)$ is given in the form

$$\Psi_{ij}(x)/\left(\sum_{t=0}^{\overline{\overline{b}}_q}\overline{\overline{\beta}}_{t,i,j,k,l,\mu,\kappa,\lambda}x^t\right)=\sum_{t=0}^{\Psi_q-\overline{\overline{b}}_q}\gamma_{t,i,j,k,l,\mu,\kappa,\lambda}x^t.$$

Finally, next equation

$$\Phi_{ij}(x)=\sum_{\mu=1}^{n}\sum_{l=1}^{\min\{\mu,r\}}\sum_{k=1}^{\min\{i,r\}}\sum_{\lambda=1}^{n}\sum_{\kappa=1}^{m}\sum_{t=0}^{\Psi_q-\overline{\overline{b}}_q+\overline{b}_q}$$

$$\left(\sum_{t_1=0}^{t}\overline{\beta}_{t_1,i,j,k,l,\mu,\kappa,\lambda}\gamma_{t-t_1,i,j,k,l,\mu,\kappa,\lambda}\right)x^t,$$

is equal to form (3.6.14). □

Remark 3.6.1. *It is notable that similar theorem can be derived by observing the second of the statements of Theorem 3.2.7.*

Algorithm 3.5 Calculating the MP-inverse application LDL^* full-rank actorization

---

Input: Rational matrix $A(x) \in \mathbf{C}(x)_r^{m\times n}$ (with elements in form (3.5.1)).

1: Apply the Algorithm 3.4 for calculating LDL^* full-rank decomposition of matrix $(A^*A)^*(A^*A)$, where $L \in \mathbf{C}(x)^{n\times r}$ and $D \in \mathbf{C}(x)^{r\times r}$ matrix with elements in the form of (3.3.2).

2: Present rational matrix $R = L^*LDL^*L$ in the following form: $R(x) = \frac{1}{p(x)}P(x)$, wherein $p(x)$ some polynomial (equal to the LCM in polynomial denominators in R) and $P(x)$ is a polynomial matrix.

3: Compute the inverse polynomial matrix P using the Algorithm 3.2 out of Ref. [74]. Generate the inverse matrix $N = R^{-1} = p(x)P^{-1}$ and transform it into the form of (3.6.1).

4: For every $i = \overline{1,m}, j = \overline{1,n}, \mu = \overline{1,n}, k = \overline{1,\min\{i,r\}}, l = \overline{1,\min\{\mu,r\}}, \kappa = \overline{1,m}, \lambda = \overline{1,n}$ execute evaluation (3.6.2)–(3.6.13).

5: For $i=\overline{1,m}$, $j=\overline{1,n}$ calculate polynomials denominators elements $A^{\dagger}_{i,j}$ on the following way

$$\Psi_{ij}(x) = LCM\{\sum_{t=0}^{\overline{\overline{b}}_q} \overline{\overline{\beta}}_{t,i,j,k,l,\mu,\kappa,\lambda} x^t \mid$$

$$k=\overline{1,\min\{i,r\}}, l=\overline{1,\min\{\mu,r\}}, \mu=\overline{1,n}, \kappa=\overline{1,m}, \lambda=\overline{1,n}\},$$

and mark with $\Psi_i(x) = \sum_{t=0}^{\Psi_q} \Psi_{t,i} x^t$.

6: For every $i=\overline{1,m}$, $\mu=\overline{1,n}$, $k=\overline{1,\min\{i,r\}}$, $l=\overline{1,\min\{\mu,r\}}$ determine the following polynomials:

$$\Psi_i(x) / \left(\sum_{t=0}^{\overline{\overline{b}}_q} \overline{\overline{\beta}}_{t,i,k,l,\mu} x^t \right),$$

and denote with $\Gamma_{i,l,k,\mu}(x) = \sum_{t=0}^{\Psi_q - \overline{\overline{b}}_q} \gamma_{t,i,k,l,\mu} x^t$.

7: For $i=\overline{1,m}$, $j=\overline{1,n}$ calculate the numerator of polynomials:

$$\Phi_{ij}(x) = \sum_{t=0}^{\Psi_q - \overline{\overline{b}}_q + \overline{b}_q} \left(\sum_{\mu=1}^{n} \sum_{l=1}^{\min\{\mu,r\}} \sum_{k=1}^{\min\{i,r\}} \sum_{\lambda=1}^{n} \sum_{\kappa=1}^{m} \sum_{t_1=0}^{t} \overline{\beta}_{t_1,i,j,k,l,\mu,\kappa,\lambda} \gamma_{t-t_1,i,k,l,\mu} \right) x^t. \tag{3.6.15}$$

8: For $i=\overline{1,m}$, $j=\overline{1,n}$ set (i,j)-th element of generalized inverse $A^{\dagger}$ on value $\Phi_{ij}(x)/\Psi_i(x)$.

Once again, let us say that the (i,j) -th element of the product matrix $(A^*A)^*(A^*A)$ can be calculated as

$$\sum_{l=1}^{n} \sum_{k=1}^{m} \sum_{k'=1}^{m} a^*_{kl} a_{ki} a^*_{k'l} a_{k'j}.$$

So, these polynomials will represent input Algorithm 3.4, used in Step 1 of the algorithm. Also, just to determine the input for Algorithm 3.2 from Ref. [74] in step 3 Algorithm 3.5.

3.6.1 Numerical Examples and Test Results

Here, we will examine Algorithm 3.4 and Algorithm 3.5 and compare some different implementations, over the next few examples. Finally, we will compare the time of execution of various algorithms and implementations on a set of random test matrix.

Example 3.6.1. Notice, symmetrical square rational matrix

$$A(x)=\begin{bmatrix} \frac{1+x^2}{-1+x} & \frac{x}{3+x^2} & 2+x \\ \frac{x}{3+x^2} & \frac{-1+x}{-2x+x^2} & 3+x \\ 2+x & 3+x & -1+x \end{bmatrix}.$$

Applying Algorithm 3.4 for $j=2$ we get

$$f_{22}(x) = \frac{(-1+x)x^2}{(1+x^2)(3+x^2)^2},\quad f_{32}=\frac{(-1+x)x(2+x)}{3+4x^2+x^4}.$$

For $j=3$ we gain the following "intermediate" results:

$$f_{33}(x)= \frac{36-198x-102x^2-192x^3-23x^4-54x^5+5x^6+8x^7+3x^8+4x^9+x^{10}}{(-1+x)(9+15x^2+2x^3+6x^4+x^6)}.$$

After these great results, the final result is slightly smaller in size after completion of simplification:

$$L(x)=\begin{bmatrix} 1 & 0 & 0 \\ \frac{(-1+x)x}{3+4x^2+x^4} & 1 & 0 \\ \frac{(-1+x)(2+x)}{1+x^2} & \frac{(-2+x)x(3+x^2)(9+5x+11x^2+3x^3+3x^4+x^5)}{-9+9x-15x^2+13x^3-4x^4+6x^5-x^6+x^7} & 1 \end{bmatrix},$$

$$D(x)=\begin{bmatrix} \frac{1+x^2}{-1+x} & 0 & 0 \\ 0 & \frac{(-1+x)(9+15x^2+2x^3+6x^4+x^6)}{(-2+x)x(1+x^2)(3+x^2)^2} & 0 \\ 0 & 0 & y \end{bmatrix}.$$

where

$$y=\frac{27-180x-126x^2-164x^3-40x^4-44x^5-2x^6+10x^7+2x^8+4x^9+x^{10}}{9-9x+15x^2-13x^3+4x^4-6x^5+x^6-x^7}$$

Next Moore–Penrose's inverse matrix A is obtained by applying Algorithm 3.5 the input matrix A:

$$A^{\dagger} = \begin{bmatrix} \frac{(3+x^2)^2(1+15x-12x^2-8x^3+3x^4+x^5)}{27-180x-126x^2-164x^3-40x^4-44x^5-2x^6+10x^7+2x^8+4x^9+x^{10}} & \frac{x(-108+66x+6x^2+16x^3+29x^4-8x^5+2x^6-2x^7-x^8)}{27-180x-126x^2-164x^3-40x^4-44x^5-2x^6+10x^7+2x^8+4x^9+x^{10}} & \frac{18-27x-6x^2+12x^3-4x^4+7x^5}{27-180x-126x^2-164x^3-40x^4-44x^5-2x^6+10x^7+2x^8+4x^9+x^{10}} \\ \frac{x(-108+66x+6x^2+16x^3+29x^4-8x^5+2x^6-2x^7-x^8)}{27-180x-126x^2-164x^3-40x^4-44x^5-2x^6+10x^7+2x^8+4x^9+x^{10}} & \frac{x(3+x^2)^2(6-x-9x^2+4x^3)}{27-180x-126x^2-164x^3-40x^4-44x^5-2x^6+10x^7+2x^8+4x^9+x^{10}} & \frac{x(-54-3x-69x^2+14x^3-26x^4+8x^5+x^7+x^8)}{27-180x-126x^2-164x^3-40x^4-44x^5-2x^6+10x^7+2x^8+4x^9+x^{10}} \\ \frac{18-27x-6x^2+12x^3-4x^4+7x^5}{27-180x-126x^2-164x^3-40x^4-44x^5-2x^6+10x^7+2x^8+4x^9+x^{10}} & \frac{x(-54-3x-69x^2+14x^3-26x^4+8x^5+x^7+x^8)}{27-180x-126x^2-164x^3-40x^4-44x^5-2x^6+10x^7+2x^8+4x^9+x^{10}} & \frac{9-9x+15x^2-13x^3+4x^4-6x^5+x^6-x^7}{27-180x-126x^2-164x^3-40x^4-44x^5-2x^6+10x^7+2x^8+4x^9+x^{10}} \end{bmatrix}.$$

Example 3.6.2. Consider the matrix $A(x)$ from Example 2.2.2 as a rational matrix. To calculate the elements of Moore–Penrose's inverse matrix A, LDL^* decomposition of matrix $(A^*A)^2$ be determined Algorithm 3.4 which is required in Step 1 of Algorithm 3.5. Thus, by applying Algorithm 3.5 the given matrix A, obtained following Moore–Penrose's inverse matrix A:

$$A^{\dagger} = \begin{bmatrix} \frac{-1+x}{-85-83x-2x^2+x^3} & \frac{9+17x}{85+83x+2x^2-x^3} & \frac{3+10x}{85+83x+2x^2-x^3} & \frac{26(1+x)}{-85-83x-2x^2+x^3} \\ \frac{9+17x}{85+83x+2x^2-x^3} & \frac{4+36x-2x^2+3x^3}{-85-83x-2x^2+x^3} & \frac{27-2x+2x^2-2x^3}{85+83x+2x^2-x^3} & \frac{21+19x-2x^2+5x^3}{85+83x+2x^2-x^3} \\ \frac{3+10x}{85+83x+2x^2-x^3} & \frac{27-2x+2x^2-2x^3}{85+83x+2x^2-x^3} & \frac{9-16x+x^2-x^3}{85+83x+2x^2-x^3} & \frac{7+7x-2x^2+3x^3}{85+83x+2x^2-x^3} \\ \frac{26(1+x)}{-85-83x-2x^2+x^3} & \frac{21+19x-2x^2+5x^3}{85+83x+2x^2-x^3} & \frac{7+7x-2x^2+3x^3}{85+83x+2x^2-x^3} & \frac{4-4x+8x^3}{-85-83x-2x^2+x^3} \end{bmatrix}.$$

Example 3.6.3. Now let's compare calculations for different time implementation of the Algorithm 3.4 in `MATHEMATICA` and `JAVA`.

For testing purposes [56], we generated a set of $n \times n$ matrix of rational density 1, with non-zero coefficients randomly selected from the interval $[-10, 10]$, where all the polynomials in the numerator and denominator of degree d. Table 3.5 illustrates the difference between the times obtained various implementations of the Algorithm 3.4 basic LDL^* decomposition rational matrix. Tests have shown that the `MATHEMATICA` implementation of the Algorithm 3.4 far more efficient as compared to other implementations. The main reason for the roads simplification executed in `MATHEMATICA`. Implementation of the basic algorithm (3.3.3) again markup as `MATH. basic algorithm` and obviously less efficient than the implementation of the Algorithm 3.4 in `JAVA` (`Java Algorithm 3.4`) to smaller dimensions. For the case of larger dimensions ($n = 10$) implementation in Java is much slower than the implementation in `MATHEMATICA`.

Example 3.6.4. To determine the effectiveness of further Algorithm 3.5, we compared several different algorithms for calculating Moore–Penrose's inverse. In Table 3.6, the average time of calculation are given, and also the application of different algorithms on two sets of test matrices from Ref. [88] are obtained (observing partial case $x = 1$).

First row in Table 3.5 contains the names of test matrices generated in Ref. [88]. Weather calculating the latter kind are obtained by applying Algorithm 3.5. Obviously, the algorithm is fast, but is not the most efficient algorithm for each set of test matrices. Also, it is clear that the Algorithm 3.5 had a better performance than the LDLGInverse from Ref. [58], from which it is derived.

3.7 Calculating $A_{T,S}^{(2)}$ Inverse LDL^* with Decomposition

Many representations of the full-rank of different types of generalized inverse been made, either for a given rank or specify the scope or core. For completeness, I shall quote the following external representation of the full-rank defined core and inverse image (rank, null space) introduced by Chen Sheng and [53].

Table 3.5. Execution Time (in seconds) Obtained Various Implementations of the Algorithm 3.4

n	5					10				
d	10	25	50	100	200	2	5	10	25	50
MATHEMATICA basic algoritam	0.22	0.7	1.76	5.66	16.91	0.11	0.38	1.04	3.26	8.71
MATHEMATICA Algorithm 3.4	0.03	0.09	0.16	0.41	1.02	0.09	0.13	0.21	0.52	1.23
Java Algorithm 3.4	0.026	0.15	0.64	2.85	12.32	0.22	0.85	4.15	31.22	115.36

Table 3.6. Average Time of Calculating Moore–Penroses Inverse (in seconds) Obtained by Different Algorithms

Test matrix	S_{10}	S_{50}	S_{100}	S_{150}	F_{10}	F_{50}	F_{100}	F_{150}
Lev.-Faddeev [29]	0.10	2.5	42.17	-	0.13	2.4	40.84	-
Partitioning [65]	0.03	1.18	8.20	74.55	0.01	0.45	2.81	15.4
LDLGInverse [58]	0.02	0.78	4.35	21.21	0.01	1.57	10.62	-
PseudoInverse [86]	0.12	2.57	40.43	-	0.08	2.2	41.34	-
Algorithm 3.5	0.01	0.44	2.48	15.91	0.02	0.60	6.24	38.01

Lemma 3.7.1. [53] *Let be given the matrix* $A \in \mathbb{C}_r^{m\times n}$ *and subspaces* $T \subseteq \mathbb{C}^n$ *dimension* $s \leq r$ *i* $S \subseteq \mathbb{C}^m$ *dimension* $m-s$. *Suppose that matrix* $M \in \mathbb{C}^{n\times m}$ *satisfies* $\mathscr{R}(M) = T, \mathscr{N}(M) = S$ *and suppose that it has full-rank decomposition* $M = FG$. *If the matrix A has* $\{2\}$*-inverse* $A_{T,S}^{(2)}$*,then, following the statements are valid:*

(1) *GAF is invertible matrix;*

(2) $A_{T,S}^{(2)} = F(GAF)^{-1}G = A_{\mathscr{R}(F),\mathscr{N}(G)}^{(2)}$.

Based on Lemma 3.7.1 and known representations from Refs. [9, 47, 48, 53, 62, 68] we state next additional representations with respect to (1.1.1)–(1.1.3). These representations characterize some classes generalized inverses.

Proposition 3.7.2. *Let* $A \in \mathbb{C}_r^{m\times n}$ *be an arbitrary matrix,* $0 < s \leq r$. *The following general representations for some classes generalized inverses are valid:*

(a) $A\{2\}_s = \{A_{\mathscr{R}(F),\mathscr{N}(G)}^{(2)} = F(GAF)^{-1}G|\quad F \in \mathbb{C}^{n\times s}, G \in \mathbb{C}^{s\times m},\ rank(GAF) = s\}$;

(b) $A\{2,4\}_s = \left\{A_{\mathscr{R}((GA)^*),\mathscr{N}(G)}^{(2)} = (GA)^*\left(GA(GA)^*\right)^{-1}G \mid G \in \mathbb{C}_s^{s\times m}\right\} = \left\{(GA)^{\dagger}G|\ GA \in \mathbb{C}_s^{s\times n}\right\}$;

(c) $A\{2,3\}_s = \left\{A_{\mathscr{R}(F),\mathscr{N}((AF)^*)}^{(2)} = F\left((AF)^*AF\right)^{-1}(AF)^*|\ F \in \mathbb{C}_s^{n\times s}\right\} = \left\{F(AF)^{\dagger},\ AF \in \mathbb{C}_s^{m\times s}\right\}$;

(d) $A\{1,2\} = A\{2\}_r$.

The Ref. [59] provides the theorem which is determined by the full-rank of external representation of the inverse, in the same sense as in Lemma 3.7.1. The following statement holds for rational matrices and based on LDL^* decomposition of the full-rank of an arbitrary Hermitian matrix M, which can be carried out in a similar manner described in Algorithm 3.4.

Theorem 3.7.3. [59] *Let's data square rational matrix $A \in \mathbb{C}(x)_r^{n\times n}$ normal rank r. If the LDL^* decomposition of the full-rank of an arbitrary Hermitian matrix $M \in \mathbb{C}(x)_s^{n\times n}$ rank $s \leq r$, wherein $L \in \mathbb{C}(x)_s^{n\times s}$, a $D \in \mathbb{C}(x)_s^{s\times s}$ is diagonal matrix, define the following set*

$$\mathscr{D}_{A,M} = \{x \in \mathbb{C} \mid nrang(M) = rank(L^*(x)A(x)L(x)) = s\}. \tag{3.7.1}$$

If the following is satisfied

$$nrang(M) = nrang(RAQ) = s, \tag{3.7.2}$$

the following statement is valid on the set $\mathscr{D}_{A,M}$:

$$A^{(2)}_{\mathscr{R}(L),\mathscr{N}(L^*)} = L(L^*AL)^{-1}L^* = A^{(2)}_{\mathscr{R}(M),\mathscr{N}(M)}. \tag{3.7.3}$$

Proof. Given that the term

$$M = LDL^* = (LD)(L^*), \tag{3.7.4}$$

represents the full-rank of the matrix decomposition M on the set $\mathscr{D}_{A,M}$, then the first identity in Eq. (3.7.3) is valid based on Lemma 3.7.1 and equation

$$LD(L^*ALD)^{-1}L^* = L(L^*AL)^{-1}L^* = A^{(2)}_{\mathscr{R}(L),\mathscr{N}(L^*)}.$$

The second equation $A^{(2)}_{\mathscr{R}(M),\mathscr{N}(M)} = A^{(2)}_{\mathscr{R}(L),\mathscr{N}(L^*)}$ is obviously satisfied on $\mathscr{D}_{A,M}$, based on equation (3.7.4). □

We now state the algorithm for calculating $A^{(2)}_{T,S}$ inverse of given matrix A, introduced in Ref. [59]

Algorithm 3.6 Calculating $A^{(2)}_{T,S}$ matrix inverse A based on LDL^* full-rank decomposition of matric M. **(Algorithm GenInvLDL)**

Input: Matrix $A \in \mathbb{C}(x)_r^{n\times n}$ regular rank r.

1: Select an arbitrary polynomial Hermitian matrix $M \in \mathbb{C}(x)^{n\times n}$ regular rank $s \leq r$.

2: Generalize LDL^* full-rank decomposition polynomial matrix M using the following Algorithm 3.4.

3: **Rezultat:** Generalized inverse of matrix $A^{(2)}_{\mathscr{R}(L),\mathscr{N}(L^*)}$ as product of matrices is given in equation (3.7.3).

The following statement from Ref. [59] provides practical criteria for design Hermitian matrix $M \in \mathbb{C}(x)^{n\times n}_s$, the purpose of calculating several types of generalized inverse matrix A.

Corollary 3.7.4. [59] *For given matrix $A \in \mathbb{C}(x)^{n\times n}_r$ regular rank r and an arbitrary Hermitian matrix $M \in \mathbb{C}(x)^{n\times n}_s$, where $s \leq r$, following the statements are valid at the meeting $\mathscr{D}_{A,M}$: for $M = A^*$ identity $A^{(2)}_{\mathscr{R}(L),\mathscr{N}(L^*)} = A^\dagger$ is satisfied; for $M = A^\sharp$ identity $A^{(2)}_{\mathscr{R}(L),\mathscr{N}(L^*)} = A^\dagger_{M,N}$ is satisfied; for $M = A$ it is valid the following identity $A^{(2)}_{\mathscr{R}(L),\mathscr{N}(L^*)} = A^\#$; for $M = A^k$, wherein $k \geq \mathrm{ind}(A)$, the following identity is valid $A^{(2)}_{\mathscr{R}(L),\mathscr{N}(L^*)} = A^\dagger$.*

Proof. Based on Theorem follows 3.7.3 and identity (1.1.1) i (1.1.2). □

On the basis of these effects, a wide rank of generalized inverse matrix $A \in \mathbb{C}(x)^{n\times n}_r$ is possible to determine, selecting appropriate Hermitian matrix M and calculating the full-rank of its decomposition. In Ref. [59] the following result is introduced in Theorem 3.7.3 for the case of polynomial matrices A and M.

Theorem 3.7.5. [59] *Let $A \in \mathbb{C}(x)^{n\times n}_r$ be regular rank polynomial matrix r. Let LDL^* be full-rank decomposition of an arbitrary Hermitian polynomial matrix $M \in \mathbb{C}(x)^{n\times n}_s$ regular rank $s \leq r$, where $L \in \mathbb{C}(x)^{n\times s}_s$ i $D \in \mathbb{C}(x)^{s\times s}_s$ are matrix in form (3.5.4) with elements of the form (3.3.2). Denote with $\mathscr{D}_{A,M}$ set as in (3.7.1) and an arbitrary (i, j)-th element of inverse matrix*

$N = (L^*AL)^{-1}$ on the following way:

$$n_{i,j}(x) = \sum_{k=0}^{\overline{n}_q} \overline{n}_{k,i,j} x^k \Big/ \sum_{k=0}^{\overline{\overline{n}}_q} \overline{\overline{n}}_{k,i,j} x^k, \tag{3.7.5}$$

If the condition (3.7.2) is satisfied, then the an arbitrary (i,j)-th element of generalized inverse $A^{(2)}_{\mathscr{R}(L),\mathscr{N}(L^*)}$ can be calculated as

$$\left(A^{(2)}_{\mathscr{R}(L),\mathscr{N}(L^*)}\right)_{ij}(x) = \frac{\overline{\Sigma}_{i,j}(x)}{\overline{\overline{\Sigma}}_{i,j}(x)},$$

for $x \in \mathscr{D}_{A,M}$, where $\overline{\Sigma}_{i,j}(x)$ and $\overline{\overline{\Sigma}}_{i,j}(x)$ polynomials of the form

$$\overline{\Sigma}_{i,j}(x) = \sum_{t=0}^{\overline{\overline{\Sigma}}_q - \overline{\overline{\gamma}}_q + \overline{\gamma}_q} \left(\sum_{k=1}^{\min\{j,s\}} \sum_{l=1}^{\min\{i,s\}} \sum_{t_1=0}^{t} \overline{\gamma}_{t_1,i,j,k,l} \sigma_{t-t_1,i,j,k,l} \right) x^t, \tag{3.7.6}$$

$$\overline{\overline{\Sigma}}_{i,j}(x) = \text{LCM}\left\{ \sum_{t=0}^{\overline{\overline{\gamma}}_q} \overline{\overline{\gamma}}_{t,l,j,k,l} x^t \,\middle|\, k = \overline{1,\min\{j,s\}}, l = \overline{1,\min\{i,s\}} \right\} \tag{3.7.7}$$

$$= \sum_{t=0}^{\overline{\overline{\Sigma}}_q} \overline{\overline{\sigma}}_{t,i,j} x^t, \tag{3.7.8}$$

where for $k = \overline{1,\min\{j,s\}}$, $l = \overline{1,\min\{i,s\}}$, values $\sigma_{t,i,j,k,l}$, $0 \le t \le \overline{\overline{\Sigma}}_q - \overline{\overline{\gamma}}_q$ coefficients of polynomial

$$\Sigma_{i,j,k,l}(x) = \frac{\overline{\overline{\Sigma}}_{i,j}(x)}{\sum_{t=0}^{\overline{\overline{\gamma}}_q} \overline{\overline{\gamma}}_{t,i,j,k,l} x^t},$$

whereby tags were used:

$$\overline{\gamma}_{t,i,j,k,l} = \sum_{t_2=0}^{t_1} \sum_{t_3=0}^{t_1-t_2} \overline{l}_{t_2,i,l} \overline{n}_{t_1-t_2-t_3,l,k} \overline{l}^*_{t_3,j,k}, \quad 0 \le t \le \overline{\gamma}_q = 2\overline{l}_q + \overline{n}_q \tag{3.7.9}$$

$$\overline{\overline{\gamma}}_{t,i,j,k,l} = \sum_{t_2=0}^{t_1} \sum_{t_3=0}^{t_1-t_2} \overline{\overline{l}}_{t_2,i,l} \overline{\overline{n}}_{t_1-t_2-t_3,l,k} \overline{\overline{l}}^*_{t_3,j,k}, \quad 0 \le t \le \overline{\overline{\gamma}}_q = 2\overline{\overline{l}}_q + \overline{\overline{n}}_q \tag{3.7.10}$$

Proof. How are the elements of the inverse matrix $N = (L^*AL)^{-1} = \{n_{i,j}(x)\}_{i,j=0}^{s}$ are determined by expression (3.7.5), then

$$(LN)_{ij}(x) = \sum_{l=1}^{\min\{i,s\}} l_{i,l}(x)n_{l,j}(x) = \sum_{l=1}^{\min\{i,s\}} \frac{\sum_{k=0}^{\overline{l}_q} \overline{l}_{k,i,l}x^k \sum_{k=0}^{\overline{n}_q} \overline{n}_{k,l,j}x^k}{\sum_{k=0}^{\overline{\overline{l}}_q} \overline{\overline{l}}_{k,i,l}x^k \sum_{k=0}^{\overline{\overline{n}}_q} \overline{\overline{n}}_{k,l,j}x^k}$$

$$= \sum_{l=1}^{\min\{i,s\}} \frac{\sum_{k=0}^{\overline{l}_q+\overline{n}_q} \left(\sum_{k_1=0}^{k} \overline{l}_{k_1,i,l}\overline{n}_{k-k_1,l,j} \right) x^k}{\sum_{k=0}^{\overline{\overline{l}}_q+\overline{\overline{n}}_q} \left(\sum_{k_1=0}^{k} \overline{\overline{l}}_{k_1,i,l}\overline{\overline{n}}_{k-k_1,l,j} \right) x^k}.$$

Accordingly, the following equalities:

$$\begin{aligned}
&(L(L^*AL)^{-1}L^*)_{ij}(x) \\
&= \sum_{k=1}^{\min\{j,s\}} (LN)_{ik}(x) \cdot (L^*)_{kj}(x) \\
&= \sum_{k=1}^{\min\{j,s\}} \sum_{l=1}^{\min\{i,s\}} \frac{\sum_{t_1=0}^{\overline{l}_q+\overline{n}_q} \left(\sum_{t_2=0}^{t_1} \overline{l}_{t_2,i,l}\overline{n}_{t_1-t_2,l,k} \right) x^{t_1}}{\sum_{t_1=0}^{\overline{\overline{l}}_q+\overline{\overline{n}}_q} \left(\sum_{t_2=0}^{t_1} \overline{\overline{l}}_{t_2,i,l}\overline{\overline{n}}_{t_1-t_2,l,k} \right) x^{t_1}} \frac{\sum_{t_2=0}^{\overline{l}_q} \overline{l}^*_{t_2,j,k}x^{t_2}}{\sum_{t_2=0}^{\overline{\overline{l}}_q} \overline{\overline{l}}^*_{t_2,j,k}x^{t_2}} \\
&= \sum_{k=1}^{\min\{j,s\}} \sum_{l=1}^{\min\{i,s\}} \frac{\sum_{t_1=0}^{2\overline{l}_q+\overline{n}_q} \left(\sum_{t_2=0}^{t_1} \sum_{t_3=0}^{t_1-t_2} \overline{l}_{t_2,i,l}\overline{n}_{t_1-t_2-t_3,l,k}\overline{l}^*_{t_3,j,k} \right) x^{t_1}}{\sum_{t_1=0}^{2\overline{\overline{l}}_q+\overline{\overline{n}}_q} \left(\sum_{t_2=0}^{t_1} \sum_{t_3=0}^{t_1-t_2} \overline{\overline{l}}_{t_2,i,l}\overline{\overline{n}}_{t_1-t_2-t_3,l,k}\overline{\overline{l}}^*_{t_3,j,k} \right) x^{t_1}}.
\end{aligned}$$

Based on Theorem 3.7.3, Eq. (3.7.3) is satisfied; $x \in \mathbb{C}_s(M)$, and an arbitrary (i,j)-th element of inverse $A^{(2)}_{\mathscr{R}(L),\mathscr{N}(L^*)}$ has form

$$\left(A^{(2)}_{\mathscr{R}(L),\mathscr{N}(L^*)}\right)_{ij} = \sum_{k=1}^{\min\{j,s\}} \sum_{l=1}^{\min\{i,s\}} \frac{\sum_{t=0}^{\overline{\gamma}_q} \overline{\gamma}_{t,i,j,k,l}x^t}{\sum_{t=0}^{\overline{\overline{\gamma}}_q} \overline{\overline{\gamma}}_{t,i,j,k,l}x^t} = \frac{\overline{\Sigma}_{i,j}(x)}{\overline{\overline{\Sigma}}_{i,j}(x)},$$

where the numerator and denominator polynomials possible to determine the following ways:

$$\overline{\overline{\Sigma}}_{i,j}(x) = \text{LCM}\left\{\sum_{t=0}^{\overline{\overline{\gamma}}_q} \overline{\overline{\gamma}}_{t,i,j,k,l}x^t \,\middle|\, k=\overline{1,\min\{j,s\}},\ l=\overline{1,\min\{i,s\}}\right\}$$

$$= \sum_{t=0}^{\overline{\overline{\Sigma}}_q} \overline{\overline{\sigma}}_{t,i}x^t,$$

$$\overline{\Sigma}_{i,j}(x) = \sum_{k=1}^{j}\sum_{l=1}^{s}\left(\Sigma_{i,j,k,l}(x)\sum_{t=0}^{\overline{\gamma}_q}\overline{\gamma}_{t,i,j,k,l}x^t\right),$$

where each polynomial $\Sigma_{i,j,k,l}(x)$ determined with

$$\Sigma_{i,j,k,l}(x) = \overline{\overline{\Sigma}}_{i,j}(x) / \sum_{t=0}^{\overline{\overline{\gamma}}_q} \overline{\overline{\gamma}}_{t,i,j,k,l}x^t = \sum_{t=0}^{\overline{\overline{\Sigma}}_q - \overline{\overline{\gamma}}_q} \sigma_{t,i,j,k,l}x^t.$$

Accordingly, each numerator polynomial $\overline{\Sigma}_{i,j}(x)$ is of thc form (3.7.6), thus completed the proof. □

Further use of the algorithm [59] is to calculate generalized inverse polynomial matrix, based on the previous theorem. It is based on LDL^* decomposition of the full-rank of appropriate polynomial matrix, using Algorithm 3.2 and calculating inverse polynomial matrix using the algorithm 3.2 out of work [74]. In order to apply Algorithm 3.2 from Ref. [74] to the rational matrix L^*AL, it should first be represented as a quotient polynomial matrices and polynomials.

Algorithm 3.7 Calculation of $A^{(2)}_{T,S}$ using the inverse matrix polynomial LDL^* factorization full-rank of arbitrary Hermitian polynomial matrix M. **(Algorithm GenInvLDL2)**

Input: Polynomial matrices $A(x) \in \mathbb{C}(x)^{n\times n}_r$ normal rank r.

1: Choose an arbitrary fixed Hermitian $n \times n$ matrix polynomial M normal rank $s \leq r$.

2: Calculate LDL^* decomposition of the full-rank of the matrix M using an Algorithm 3.2, where are the matrix elements of a rational L take the form of (3.3.2)

3: Calculate the lowest common denominators of the rational polynomial matrix L^*AL, so that the following equation valid: $L^*AL = \frac{1}{p(x)}P(x)$, where $p(x)$ and the corresponding polynomial $P(x)$ is a polynomial matrix.

4: Compute the inverse matrix $P^{-1}(x)$ using Algorithm 3.2 out of work [74]. Determine the inverse matrix $N = (L^*AL)^{-1}$ as the product of $p(x) \cdot P^{-1}(x)$, where the elements of the matrix N form (3.7.5).

5: Introduce sustitutions $\overline{\gamma}_q = 2\overline{l}_q + \overline{n}_q$, $\overline{\overline{\gamma}}_q = 2\overline{\overline{l}}_q + \overline{\overline{n}}_q$, and for $i, j = \overline{1,n}$ perform Steps 5.1–5.5.

5.1: For $k = \overline{1, \min\{j,s\}}$, $l = \overline{1, \min\{i,s\}}$, make the following calculation:

$$\begin{aligned} \overline{\gamma}_{t,i,j,k,l} &= \sum_{t_2=0}^{t_1} \sum_{t_3=0}^{t_1-t_2} \overline{l}_{t_2,i,l} \overline{n}_{t_1-t_2-t_3,l,k} \overline{l}^*_{t_3,j,k}, \quad 0 \le t \le \overline{\gamma}_q, \\ \overline{\overline{\gamma}}_{t,i,j,k,l} &= \sum_{t_2=0}^{t_1} \sum_{t_3=0}^{t_1-t_2} \overline{l}_{t_2,i,l} \overline{\overline{n}}_{t_1-t_2-t_3,l,k} \overline{\overline{l}}^*_{t_3,j,k}, \quad 0 \le t \le \overline{\overline{\gamma}}_q. \end{aligned}$$

5.2: Calculate the numerator polynomial (i, j) – this element matrix $A^{(2)}_{\mathscr{R}(L),\mathscr{N}(L^*)}$ as

$$\text{LCM}\left\{ \sum_{t=0}^{\overline{\overline{\gamma}}_q} \overline{\overline{\gamma}}_{t,i,j,k,l} x^t \,\middle|\, k = \overline{1, \min\{j,s\}},\ l = \overline{1, \min\{i,s\}} \right\},$$

and denote with $\overline{\overline{\Sigma}}_{i,j}(x) = \sum_{t=0}^{\overline{\overline{\Sigma}}_q} \overline{\overline{\sigma}}_{t,i,j} x^t$.

5.3: For $k = \overline{1, \min\{j,s\}}$, $l = \overline{1, \min\{i,s\}}$ calculate polynomial $\overline{\overline{\Sigma}}_{i,j}(x) / \sum_{t=0}^{\overline{\overline{\gamma}}_q} \overline{\overline{\gamma}}_{t,i,j,k,l} x^t$ and denote with $\Sigma_{i,j,k,l}(x) = \sum_{t=0}^{\overline{\overline{\Sigma}}_q - \overline{\overline{\gamma}}_q} \sigma_{t,i,j,k,l} x^t$.

5.4: Calculate the numerator polynomial (i, j) – this element matrix $A^{(2)}_{\mathscr{R}(L),\mathscr{N}(L^*)}$ as

$$\overline{\Sigma}_{i,j}(x) = \sum_{t=0}^{\overline{\overline{\Sigma}}_q - \overline{\overline{\gamma}}_q + \overline{\gamma}_q} \left(\sum_{k=1}^{\min\{j,s\}} \sum_{l=1}^{\min\{i,s\}} \sum_{t_1=0}^{t} \overline{\gamma}_{t_1,i,j,k,l}\, \sigma_{t-t_1,i,j,k,l} \right) x^t,$$

5.5: **Input:** (i, j)-th element of inverse matrix $A^{(2)}_{\mathscr{R}(L),\mathscr{N}(L^*)}$ is given in form $\overline{\Sigma}_{i,j}(x)/\overline{\overline{\Sigma}}_{i,j}(x)$.

3.7.1 Numerical Examples

In the next few examples we describe the algorithm and test several implementations like comparing runtime papers random matrices and test matrices from Ref. [88].

Example 3.7.1. Consider the polynomial matrix $A(x)$ given in Example 2.2.2. As the A is a real symmetric matrix, its Moore–Penrose's inverse can be calculated using the algorithm GenInvLDL2 the matrix $M = A = A^*$. After several calculations required in Step 5 above algorithm, obtained the following result:

$A^\dagger = A^\# =$

$$\begin{bmatrix} \frac{-1+x}{-85-83x-2x^2+x^3} & \frac{9+17x}{85+83x+2x^2-x^3} & \frac{3+10x}{85+83x+2x^2-x^3} & \frac{26(1+x)}{-85-83x-2x^2+x^3} \\ \frac{9+17x}{85+83x+2x^2-x^3} & \frac{4+36x-2x^2+3x^3}{-85-83x-2x^2+x^3} & \frac{27-2x+2x^2-2x^3}{85+83x+2x^2-x^3} & \frac{21+19x-2x^2+5x^3}{85+83x+2x^2-x^3} \\ \frac{3+10x}{85+83x+2x^2-x^3} & \frac{27-2x+2x^2-2x^3}{85+83x+2x^2-x^3} & \frac{9-16x+x^2-x^3}{85+83x+2x^2-x^3} & \frac{7+7x-2x^2+3x^3}{85+83x+2x^2-x^3} \\ \frac{26(1+x)}{-85-83x-2x^2+x^3} & \frac{21+19x-2x^2+5x^3}{85+83x+2x^2-x^3} & \frac{7+7x-2x^2+3x^3}{85+83x+2x^2-x^3} & \frac{4-4x+8x^3}{-85-83x-2x^2+x^3} \end{bmatrix}.$$

Example 3.7.2. Consider the symmetric matrix F_6 generated in Ref. [88] according to the variable x, i.e.,

$$F_6 = \begin{bmatrix} 6+x & 5+x & 4+x & 3+x & 2+x & 1+x \\ 5+x & 5+x & 4+x & 3+x & 2+x & 1+x \\ 4+x & 4+x & 4+x & 3+x & 2+x & 1+x \\ 3+x & 3+x & 3+x & 3+x & 2+x & 1+x \\ 2+x & 2+x & 2+x & 2+x & 1+x & x \\ 1+x & 1+x & 1+x & 1+x & x & -1+x \end{bmatrix}.$$

Now, in order to determine the Moore–Penrose's inverse matrix's F_6, consider the polynomial matrix $M = F_6^* = F_6$. How is the matrix F_6 equal to 5 matrix L and D received LDL^* matrix factorization F_6 have the following form:

$$L = \begin{bmatrix} 1 & 0 & 0 & 0 & 0 \\ \frac{5+x}{6+x} & 1 & 0 & 0 & 0 \\ \frac{4+x}{6+x} & \frac{4+x}{5+x} & 1 & 0 & 0 \\ \frac{3+x}{6+x} & \frac{3+x}{5+x} & \frac{3+x}{4+x} & 1 & 0 \\ \frac{2+x}{6+x} & \frac{2+x}{5+x} & \frac{2+x}{4+x} & \frac{2+x}{3+x} & 1 \\ \frac{1+x}{6+x} & \frac{1+x}{5+x} & \frac{1+x}{4+x} & \frac{1+x}{3+x} & 2 \end{bmatrix},$$

$$D = \begin{bmatrix} 6+x & 0 & 0 & 0 & 0 \\ 0 & \frac{5+x}{6+x} & 0 & 0 & 0 \\ 0 & 0 & \frac{4+x}{5+x} & 0 & 0 \\ 0 & 0 & 0 & \frac{3+x}{4+x} & 0 \\ 0 & 0 & 0 & 0 & -\frac{1}{3+x} \end{bmatrix}.$$

Therefore, inverse matrix $(L^* F_6 L)^{-1}$ is 5×5 rational matrix in the following form

$$\begin{bmatrix} 1 & \frac{-11-2x}{6+x} & \frac{4+x}{5+x} & 0 & 0 \\ \frac{-11-2x}{6+x} & \frac{157+56x+5x^2}{36+12x+x^2} & \frac{-98-40x-4x^2}{30+11x+x^2} & \frac{3+x}{4+x} & 0 \\ \frac{4+x}{5+x} & \frac{-98-40x-4x^2}{30+11x+x^2} & \frac{122+54x+6x^2}{25+10x+x^2} & \frac{-352-183x-23x^2}{120+54x+6x^2} & \frac{4+3x}{18+6x} \\ 0 & \frac{3+x}{4+x} & \frac{-352-183x-23x^2}{120+54x+6x^2} & \frac{1816+932x+88x^2-9x^3}{576+288x+36x^2} & \frac{-104-62x+24x^2+9x^3}{432+252x+36x^2} \\ 0 & 0 & \frac{4+3x}{18+6x} & \frac{-104-62x+24x^2+9x^3}{432+252x+36x^2} & \frac{-44-60x-36x^2-9x^3}{324+216x+36x^2} \end{bmatrix},$$

and, finally, was generated following Moore–Penrose's inverse backing last step of the algorithm GenInvLDL2:

$$F_6^{\dagger} = \begin{bmatrix} 1 & -1 & 0 & 0 & 0 & 0 \\ -1 & 2 & -1 & 0 & 0 & 0 \\ 0 & -1 & 2 & -\frac{5}{6} & -\frac{1}{3} & \frac{1}{6} \\ 0 & 0 & -\frac{5}{6} & \frac{1}{36}(28-9x) & \frac{4}{9} & \frac{1}{36}(4+9x) \\ 0 & 0 & -\frac{1}{3} & \frac{4}{9} & \frac{1}{9} & -\frac{2}{9} \\ 0 & 0 & \frac{1}{6} & \frac{1}{36}(4+9x) & -\frac{2}{9} & \frac{1}{36}(-20-9x) \end{bmatrix}.$$

To calculate the inverse matrix Drazin F_6, index $k \geq ind(A)$ be determined. Accordingly, for $k=2$ matrix factorization should determine $M = F_6^2$, and the result obtained is equal to Moore–Penrose's inverse, i.e., $F_6^D = F_6^{\dagger} = F_6^{\#}$.

Example 3.7.3. Notice, now 6×6 matrix A_2 rank 2 generated in Ref. [88], given in the form of a polynomial

$$A_2 = \begin{bmatrix} x+10 & x+9 & x+8 & x+7 & x+6 & x+5 \\ x+9 & x+8 & x+7 & x+6 & x+5 & x+4 \\ x+8 & x+7 & x+6 & x+5 & x+4 & x+3 \\ x+7 & x+6 & x+5 & x+4 & x+3 & x+2 \\ x+6 & x+5 & x+4 & x+3 & x+2 & x+1 \\ x+5 & x+4 & x+3 & x+2 & x+1 & x \end{bmatrix}.$$

Using Algorithm 3.2, for $j=2$ we get the following inter-results:

$$f_{2,2}(x) = \frac{(9+x)^2}{10+x}, \; f_{3,2}(x) = \frac{(8+x)(9+x)}{10+x},$$
$$f_{4,2}(x) = \frac{(7+x)(9+x)}{10+x}, f_{5,2}(x) = \frac{(6+x)(9+x)}{10+x},$$
$$f_{6,2}(x) = \frac{(5+x)(9+x)}{10+x}.$$

Accordingly, we have the following rational matrix included in the LDL^* decomposition of the starting matrix

$$L = \begin{bmatrix} 1 & 0 \\ \frac{9+x}{10+x} & 1 \\ \frac{8+x}{10+x} & 2 \\ \frac{7+x}{10+x} & 3 \\ \frac{6+x}{10+x} & 4 \\ \frac{5+x}{10+x} & 5 \end{bmatrix}, \quad D = \begin{bmatrix} 10+x & 0 \\ 0 & -\frac{1}{10+x} \end{bmatrix}.$$

Applying Algorithm GenInvLDL2 and results of the Equation 2.6.3, the consequences were obtained following each other equal to General inverses:

$$A_2^{\dagger} = A_2^{\#} = A^D$$

$$= \begin{bmatrix} \frac{1}{147}(-8-3x) & \frac{1}{735}(-17-9x) & \frac{2-x}{245} & \frac{1}{735}(29+3x) & \frac{1}{735}(52+9x) & \frac{5+x}{49} \\ \frac{1}{735}(-17-9x) & \frac{-10-9x}{1225} & \frac{25-9x}{3675} & \frac{80+9x}{3675} & \frac{9(5+x)}{1225} & \frac{1}{735}(38+9x) \\ \frac{2-x}{245} & \frac{25-9x}{3675} & \frac{20-3x}{3675} & \frac{5+x}{1225} & \frac{10+9x}{3675} & \frac{1}{735}(1+3x) \\ \frac{1}{735}(29+3x) & \frac{80+9x}{3675} & \frac{5+x}{1225} & \frac{-50-3x}{3675} & \frac{-115-9x}{3675} & \frac{1}{245}(-12-x) \\ \frac{1}{735}(52+9x) & \frac{9(5+x)}{1225} & \frac{10+9x}{3675} & \frac{-115-9x}{3675} & \frac{-80-9x}{1225} & \frac{1}{735}(-73-9x) \\ \frac{5+x}{49} & \frac{1}{735}(38+9x) & \frac{1}{735}(1+3x) & \frac{1}{245}(-12-x) & \frac{1}{735}(-73-9x) & \frac{1}{147}(-22-3x) \end{bmatrix}.$$

Example 3.7.4. Denote the square symmetrical matrix

$$A(x) = \begin{bmatrix} 2+3x^{1000} & 1-x^{1000} & 2+x & -3+x^{1000} \\ 1-x^{1000} & -1+x^{1000} & 3+2x & 3x^{1000} \\ 2+x & 3+2x & -1+x & 1+2x \\ -3+x^{1000} & 3x^{1000} & 1+2x & x^{1000} \end{bmatrix}$$

By applying this input in `MATHEMATICA` the following result is attained:

$$L(x) = \begin{bmatrix} 1 & 0 & 0 & 0 \\ \frac{1-x^{1000}}{2+3x^{1000}} & 1 & 0 & 0 \\ \frac{2+x}{2+3x^{1000}} & \frac{4+3x+11x^{1000}+7x^{1001}}{-3+x^{1000}+2x^{2000}} & 1 & 0 \\ \frac{-3+x^{1000}}{2+3x^{1000}} & \frac{3+2x^{1000}+10x^{2000}}{-3+x^{1000}+2x^{2000}} & -y & 1 \end{bmatrix},$$

$$D(x) = \begin{bmatrix} 2+3x^{1000} & 0 & 0 & 0 \\ 0 & \frac{-3+x^{1000}+2x^{2000}}{2+3x^{1000}} & 0 & 0 \\ 0 & 0 & -m & 0 \\ 0 & 0 & 0 & -n \end{bmatrix}. \quad \square$$

$$where, y = \frac{-18-15x+9x^{1000}+5x^{1001}-36x^{2000}-20x^{2001}}{-1+9x+3x^2+44x^{1000}+53x^{1001}+17x^{1002}+2x^{2000}-2x^{2001}}$$

$$m = \frac{-1+9x+3x^2+44x^{1000}+53x^{1001}+17x^{1002}+2x^{2000}-2x^{2001}}{-3+x^{1000}+2x^{2000}}$$

$$n = \frac{105+207x+84x^2+71x^{1000}-30x^{1001}-7x^{1002}+26x^{2000}+158x^{2001}+72x^{2002}+32x^{3000}-32x^{3001}}{-1+9x+3x^2+44x^{1000}+53x^{1001}+17x^{1002}+2x^{2000}-2x^{2001}}$$

Notice that coefficients of matrix $A(x)$ from this example, based on form (3.5.1), are as follows:

$$\begin{aligned}
A_0 &= \{\{\{2,1,2,-3\},\{1,-1,3,0\},\{2,3,-1,1\},\{-3,0,1,0\}\},\\
A_1 &= \{\{0,0,1,0\},\{0,0,2,0\},\{1,2,1,2\},\{0,0,2,0\}\},\\
A_2 &= \ldots = A_{999} = \mathbf{0},\\
A_{1000} &= \{\{3,-1,0,1\},\{-1,1,0,3\},\{0,0,0,0\},\{1,3,0,1\}\}\}
\end{aligned}$$

It is clear that benefits using sparsity structures in examples like this one are significant.

3.7.2 Implementation Details and Results of Testing

In this section, the implementation Algorithm 3.10, called GenInvLDL2 in the software package `MATHEMATICA`. Function GenInvLDL2 includes two parameters: the given polynomial matrix A and an arbitrary fixed matrix W. Depending on the selection of the matrix W, we get the corresponding generalized inverse matrix A.

```
LDLATS2[A_List, W_List] :=
  Module[{N1, L, D, i, j, k, l, n, m, r1, r2, r3, p, q, f, Num, Den, s = MatrixRank[W]},
   {n, m} = Dimensions[W];
   {L, D} = PolynomialLDLDecomposition[W];
   N1 =
    ExpandNumerator[ExpandDenominator[Together[Simplify[Inverse[Transpose[L].A.L]]]]];
   f = Num = Den = Table[0, {n}, {m}];
   For[i = 1, i ≤ n, i++,
    For[j = 1, j ≤ m, j++,
      For[k = 1, k ≤ Min[j, s], k++,
       For[l = 1, l ≤ Min[i, s], l++,
         Num[[k, l]] = 0; Den[[k, l]] = 0;
         For[r1 = 0, r1 ≤
            Max[Exponent[Numerator[L[[i, l]]], x], Exponent[Denominator[L[[i, l]]], x]] +
             Max[Exponent[Numerator[N1[[l, k]]], x], Exponent[Denominator[N1[[l, k]]], x]] +
             Max[Exponent[Numerator[L[[j, k]]], x], Exponent[Denominator[L[[j, k]]], x]],
            r1++,
          p = ∑_{r2=0}^{r1} ∑_{r3=0}^{r1-r2} Coefficient[Numerator[L[[i, l]]], x, r2] *
              Coefficient[Numerator[N1[[l, k]]], x, r1 - r2 - r3] *
              Conjugate[Coefficient[Numerator[L[[j, k]]], x, r3]];
          q = ∑_{r2=0}^{r1} ∑_{r3=0}^{r1-r2} Coefficient[Denominator[L[[i, l]]], x, r2] *
              Coefficient[Denominator[N1[[l, k]]], x, r1 - r2 - r3] *
              Conjugate[Coefficient[Denominator[L[[j, k]]], x, r3]];
          Num[[k, l]] += p * x^r1;
          Den[[k, l]] += q * x^r1;
          If[Den[[k, l]] == 0, Num[[k, l]] = 0; Den[[k, l]] = 1];
         ];];];
      f[[i, j]] = Together[Simplify[∑_{k=1}^{Min[j,s]} ∑_{l=1}^{Min[i,s]} Simplify[Num[[k, l]] / Den[[k, l]]]]];
     ];];
   Return[f];
  ];
```

Implementation of Algorithm 3.6 (GenInvLDL) is extremely simple, and is given the following code, assuming that the matrix equation (3.7.3) solved by using MATHEMATICA function Inverse.

```
ATS2Inverse[A_List, W_List] :=
  Module[{N1, L, D},
   {L, D} = PolynomialLDLDecomposition[W];
   N1 = Inverse[Transpose[L].A.L] // Simplify;
   Return[Simplify[L.N1.Transpose[L]]];
  ];
```

Example 3.7.5. GenInvLDL now examine the efficiency of the algorithm, comparing a group of different algorithms calculation of Moore–Penrose's inverse matrix. The Table 3.6 provides the mean time calculation obtained by applying these algorithms to test three sets of Hermitian matrices of [88], observing the partial case $x = 1$.

First row of Table 3.7 contains the names of the test matrix, where they observed three groups of matrix from Ref. [88]. The times obtained GenInvLDL algorithm are listed in the last row. Obviously, this is not the most efficient algorithm for each set of numerical matrix, but it is on par with the algorithm that is introduced in Courrieu [11]. Long-term weather birthdates listed dashes.

Example 3.7.6. The following comparative experiments, a few random Hermitian polynomial matrix $A \in \mathbb{C}(x)_r^{n \times n}$ different size and density were observed. Thus, the greatest net value of the test matrix A is denoted as k_r, where the smallest non-zero net value of the matrix A is equal to 1 (similar tests were carried out in the Ref. [11]. Some of the obtained times are illustrated in Table 3.8, giving an insight into the effectiveness and GenInvLDL GenInvLDL2 algorithms. Obviously, GenInvLDL2 algorithm is slightly more efficient, given that specializes in polynomial matrices.

Some very efficient method for generalized inverses evaluation are numerical unstable when matrix is non-singular, but ill-conditioned (see definition in Ref. [12]). Off course, rank deficient matrices are processed faster than full-rank matrices same size, what is direct result smaller matrix sizes L and D, produced by Algorithm 2.2.1. However, processing times can grow exponentially with matrix size and density increment. Computation generalized inverses a symbolical matrix is computationally expensive ($O(n^3)$ problem), and is sensitive to ill-conditioned matrices, off course.

3.8 Representations Complete Ranking Based on the *QR* Decomposition

Let us mention first the basic concepts regarding *QR* factorization matrices.

Table 3.7. Average Time of Calculation (in seconds) Obtained by Different Algorithms and GenInvLDL Algorithm

Test matrix	A_{10}	A_{50}	A_{100}	A_{150}	S_{10}	S_{50}	S_{100}	S_{150}	F_{10}	F_{50}	F_{100}	F_{150}
Partitioning [65]	0.01	0.32	2.45	12.25	0.02	1.14	7.20	72.35	0.01	0.35	2.61	13.1
Lev.-Faddeev [29]	-	0.11	2.31	42.323	-	0.12	2.62	41.84	-	0.03	2.7	42.72
PseudoInverse [86]	0.03	1.46	22.31	-	0.12	2.57	40.22	-	0.09	2.23	31.37	-
Courrieu [11]	0.02	0.34	5.54	34.22	0.01	0.24	2.56	14.2	0.01	0.60	5.32	35.11
LDLGInverse [58]	0.02	1.33	10.64	-	0.02	0.48	4.15	21.21	0.01	1.56	11.44	-
GenInvLDL	0.02	0.65	5.05	41.28	0.01	0.33	2.56	15.11	0.02	0.81	6.03	36.01

Table 3.8. Average Time of Calculation (in seconds) for the Randomly Generated Polynomial Matrix

n	128			256			512		
k_r	16	256	4096	16	256	4096	16	256	4096
GenInvLDL full-rank	4.11	4.23	4.44	24.01	25.24	26.31	222.76	225.90	226.59
GenInvLDL rank-deficient	3.04	3.75	3.86	25.1	25.24	25.53	185.90	190.36	190.46
GenInvLDL2 full-rank	3.59	3.72	3.91	25.11	27.48	28.52	214.8	216.8	221.81
GenInvLDL2 rank-deficient	2.4	2.43	2.82	22.1	22.7	23.3	165.07	168.39	171.2

Theorem 3.8.1. *If $A \in \mathbb{R}_r^{m\times n}$, then exists an orthogonal matrix $Q \in \mathbb{R}^{m\times n}$ and upper triangular $R \in \mathbb{R}^{n\times n}$, such that $A = QR$.*

Theorem 3.8.2. *If $A = QR$ represents the QR factorization of the matrix $A \in \mathbb{C}_r^{m\times n}$, then*

$$A^{\dagger} = R^{\dagger}Q^* = R^*(RR^*)^{-1}Q^*.$$

Using a full-rank factorization generalized inverse and singular value decomposition to calculate the full-rank factorization, is obtained following the presentation of generalized inverse from [62].

Statement 3.8.1. *If $A = QR$ QR-factorization from $A \in \mathbb{C}_r^{n\times n}$, where $Q^*Q = I_r$ i R upper trapezoidal, then :*

$X \in A\{2\} \Leftrightarrow X = Y(ZQRY)^{-1}Z, rank(ZQRY) = t$, *where* $Y \in \mathbb{C}^{n\times t}, Z \in \mathbb{C}^{t\times m}, t \in \{1,\ldots,r\}$;
In case of $Y \in \mathbb{C}^{n\times r}$ *i* $Z \in \mathbb{C}^{r\times m}$ *dobijamo*
$X \in A\{1,2\} \Leftrightarrow X = Y(ZQRY)^{-1}Z, rank(ZQRY) = r$;
$X \in A\{1,2,3\} \Leftrightarrow X = Y(RY)^{-1}L^*E, rank(RY) = r$;
$X \in A\{1,2,4\} \Leftrightarrow X = R^*(ZQRR^*)^{-1}Z, rank(ZQRR^*) = r$;
$A^{\dagger} = R^*(RR^*)^{-1}Q^* = R^{\dagger}Q^*$
$A^{\sharp}$ *exists if and only if RQ invertible, i* $A^{\#} = Q(RQ)^{-2}R$.

Of the standard premium features `QRDecomposition`, `QRDecomposition[m]` gives *QR* decomposition numerical matrix *M*. The result is a list of $\{q, r\}$, where *q* orthogonal matrix, and *r* is upper triangular matrix. The starting matrix *m* is equal to `Conjugate[Transpose[q]].r..` Now calculate Moore–Penrose using inverse *QR* decomposition can be implemented with the following function:

```
QRMP[a_]:=
```

```
Block[{b=a,q,r},
{q,r}=QRDecomposition[b];
Return[Hermit[r].Inverse[r.Hermit[r]].q];
]
```

Example 3.8.1.

```
In[1]:= a={{1, 2, 3}, {3, 2, 1.}}
In[2]:= {q,r}=QRDecomposition[a]
Out[2]= {{{-0.316228, -0.948683}, {0.948683,
-0.316228}},> {{-3.16228, -2.52982, -1.89737}, {0,
1.26491, 2.52982}}}
In[3]:=Transpose[q].r
Out[3]= {{1., 2., 3.}, {3., 2., 1.}}
In[4]:= x=Transpose[r].Inverse[r.Transpose[r]].q
Out[5]= {{-0.166667, 0.333333}, {0.0833333,
0.0833333},
> {0.333333, -0.166667}}
```

Equations (1)–(4) can be checked:

$$axa = \begin{bmatrix} 1. & 2. & 3. \\ 3. & 2. & 1. \end{bmatrix}, \quad xax = \begin{bmatrix} -0.166667 & 0.333333 \\ 0.0833333 & 0.0833333 \\ 0.333333 & -0.166667 \end{bmatrix}$$

$$ax = \begin{bmatrix} 1. & -3.88578 * 10^{-16} \\ 2.77556 * 10^{-16} & 1. \end{bmatrix},$$

$$xa = \begin{bmatrix} 0.833333 & 0.333333 & -0.166667 \\ 0.333333 & 0.333333 & 0.333333 \\ -0.166667 & 0.333333 & 0.833333 \end{bmatrix}$$

Representations of Moore–Penrose's inverse $A^{\dagger}$ based on *QR* decomposition of the matrix *A* are known in the literature (see for example [33, 80]). Numerical testing of these representations is presented in Ref.

[33]. Numerical methods for calculating the least squares solutions $A^{\dagger}b$ minimum standards for linear equations $Ax = b$, based on QR factorization of the matrix A^* is investigated in Ref. [25]. This method can be used for the calculation of Moore–Penrose's inverse $A^{\dagger}b$ taking successive values of $b = e_i$, where e_i i-th unit vector for each $i = 1, \ldots, n$.

External general inverses defined core and image are very important in the theory of matrices. $\{2\}$-inverses are used in iterative methods for solving nonlinear equations [3, 43] and statistics [22, 27]. In particular, the outer inverses play an important role in the stable approximation lose caused by the problem, as well as in linear and nonlinear problems with the rank-deficient general inverses [42, 87]. On the other hand, it is well-known that the Moore–Penrose's inverse and weighted Moore–Penrose's inverse $A^{\dagger}, A^{\dagger}_{M,N}$, Drazin's group and the inverse of the $D^A, A^{\#}$, as well as Bott-Duffin's inverse $A^{(-1)}_{(L)}$ and generalized Bott-Duffin's inverse $A^{(\dagger)}_{(L)}$ be presented in a unique way as a generalized inverse $A^{(2)}_{T,S}$ for the appropriate selection of matrix T and S.

There are a large number of "full–rank" external representation of the inverse and defined ranking Foreign inverses given core and images. Here we use a representation of [53] Lemma 3.7.1 to perform subsequent results.

Representation of $A^{(2)}_{T,S}$ inversion data in Lemma 3.7.1 is very general result. This representation was used in the same study to determine representations of determinants $A^{(2)}_{T,S}$ inverses. Determinant's representation is not efficient numerical procedure for calculating the generalized inverse. Moreover, the authors of Ref. [53] not considered various possibilities for the corresponding complete factorization rank.

Here are two objectives that we stated in the Ref. [60]. First is made effective method to calculate the inverse of external application of a general representation of the full-rank. Further described are analogous algorithms to calculate $\{2,4\}$ and $\{2,3\}$-inverse, as well as two special subset of the set of external inverse.

Based on Proposition 3.7.1 and known representations from Refs. [3, 48, 53, 62, 68], in the work [60] is listed with the following betting representations (1.1.1)–(1.1.3). These representations are characterized class $\{2\}$, $\{2,4\}$ and $\{2,3\}$ of generalized inverse of a given rank.

Proposition 3.8.3. [60] *Let $A \in \mathbb{C}_r^{m\times n}$ specified matrix and let $0 < s \leq r$ be valid. Then, the following general representation for a class of generalized inverse:*

(a) $A\{2\}_s = \{A^{(2)}_{\mathscr{R}(F),\mathscr{N}(G)} = F(GAF)^{-1}G \mid F \in \mathbb{C}^{n\times s}, G \in \mathbb{C}^{s\times m}, rank(GAF) = s\}$

(b) $A\{2,4\}_s = \left\{A^{(2,4)}_{\mathscr{R}((GA)^*),\mathscr{N}(G)} = (GA)^*\left(GA(GA)^*\right)^{-1} G \mid G \in \mathbb{C}_s^{s\times m}\right\}$
$= \left\{(GA)^{\dagger} G \mid GA \in \mathbb{C}_s^{s\times n}\right\};$

(c) $A\{2,3\}_s = \left\{A^{(2,3)}_{\mathscr{R}(F),\mathscr{N}((AF)^*)} = F\left((AF)^* AF\right)^{-1}(AF)^* \mid F \in \mathbb{C}_s^{n\times s}\right\}$
$= \left\{F(AF)^{\dagger} \mid AF \in \mathbb{C}_s^{m\times s}\right\};$

(d) $A\{1,2\} = A\{2\}_r$.

This representation is based on the QR decomposition defined in Theorems 3.3.11 of Ref. [82]. Analogous QR decomposition of a complex matrix was used in Ref. [20].

Lemma 3.8.4. [60] *Let $A \in \mathbb{C}_r^{m\times n}$ be given matrix. Consider an arbitrary matrix $W \in \mathbb{C}_s^{n\times m}$, where $s \leq r$. Let QR be factorization of matrix W given in the following form*

$$WP = QR, \tag{3.8.1}$$

where P is some $m \times m$ permutation matrix, $Q \in \mathbb{C}^{n\times n}$, $Q^*Q = I_n$ a $R \in \mathbb{C}_s^{n\times m}$ the upper trapezoidal matrix. Suppose that the matrix P chosen so that Q and R can be represented as

$$Q = \begin{bmatrix} Q_1 & Q_2 \end{bmatrix}, \quad R = \begin{bmatrix} R_{11} & R_{12} \\ O & O \end{bmatrix} = \begin{bmatrix} R_1 \\ O \end{bmatrix}, \tag{3.8.2}$$

where Q_1 contains the first s columns of a matrix Q $R_{11} \in \mathbb{C}^{s\times s}$ is nonsingular.

If matrix A has {2}-inverse $A^{(2)}_{\mathscr{R}(W),\mathscr{N}(W)}$, then:

(a) $R_1P^*AQ_1$ *is invertible matrix;*

(b) $A^{(2)}_{\mathscr{R}(W),\mathscr{N}(W)} = Q_1(R_1P^*AQ_1)^{-1}R_1P^*$;

(c) $A^{(2)}_{\mathscr{R}(W),\mathscr{N}(W)} = A^{(2)}_{\mathscr{R}(Q_1),\mathscr{N}(R_1P^*)}$;

(d) $A^{(2)}_{\mathscr{R}(W),\mathscr{N}(W)} = Q_1(Q_1^*WAQ_1)^{-1}Q_1^*W$;

(e) $A^{(2)}_{\mathscr{R}(W),\mathscr{N}(W)} \in A\{2\}_s$.

Proof. (a) On the basis of assumptions is incurred

$$W = QRP^*. \tag{3.8.3}$$

It is obvious that non-trivial part of the QR decomposition (3.8.1), defined as

$$W = Q_1(R_1P^*), \tag{3.8.4}$$

actually complete factorization rank of a matrix W (see [80]). Given that A has {2}-inverse $A^{(2)}_{\mathscr{R}(W),\mathscr{N}(W)}$, then, based on the part (1) of Lemma 3.7.1 conclude that the $R_1P^*AQ_1$ invertible.

(b), (c) Based on part (2) Lemma 3.7.1, we can easily get the representation

$$A^{(2)}_{\mathscr{R}(W),\mathscr{N}(W)} = Q_1(R_1P^*AQ_1)^{-1}R_1P^*$$

the outside inverse $A^{(2)}_{\mathscr{R}(W),\mathscr{N}(W)}$ given with picture $\mathscr{R}(W) = \mathscr{R}(Q_1)$ and kernel $\mathscr{N}(W) = \mathscr{N}(R_1P^*)$.

(d) Moreover, on the basis of (3.8.4) we have

$$R_1P^* = Q_1^*W$$

and

$$Q_1(R_1P^*AQ_1)^{-1}R_1P^* = Q_1(Q_1^*WAQ_1)^{-1}Q_1^*W.$$

(e) This part of the statements follows from state 3.8.3. □

Using results from the state 3.8.3 and taking into account the equation (1.1.1)–(1.1.3), we get the following "full–rank" representations for different Foreign generalized inverses:

Corollary 3.8.5. [60] *For given matrix* $A \in \mathbb{C}_r^{m\times n}$ *and selected matrix* $W \in \mathbb{C}_s^{n\times m}$, $s \le r$, *with the representation of the full-rank* (3.8.4) *resulting from its QR decomposition* (3.8.1) *following the statements are valid:*

$$A^{(2)}_{\mathscr{R}(Q_1),\mathscr{N}(R_1P^*)} = \begin{cases} A^{\dagger}, & W = A^*; \\ A^{\dagger}_{M,N}, & W = A^{\sharp}; \\ A^{\#}, & W = A; \\ A^{D}, & W = A^k,\ k \ge \operatorname{ind}(A); \\ A^{(-1)}_{(L)}, & \mathscr{R}(W) = L,\ \mathscr{N}(W) = L^{\perp}; \\ A^{(\dagger)}_{(L)}, & \mathscr{R}(W) = S,\ \mathscr{N}(W) = S^{\perp} \end{cases} \tag{3.8.5}$$

Theorem 3.8.6. *Let* $A \in \mathbb{C}_r^{m\times n}$ *be given matrix,* $s \le r$ *be a given integer and matrices* F, G *are chosen as in Proposition 3.8.3.*

(a) *If* (3.8.4) *is full-rank factorization* $W = (GA)^*G \in \mathbb{C}_s^{n\times m}$ *then the following is valid:*

$$A^{(2)}_{\mathscr{R}(Q_1),\mathscr{N}(R_1P^*)} = Q_1(R_1P^*AQ_1)^{-1}R_1P^* \in A\{2,4\}_s. \tag{3.8.6}$$

(b) *If* (3.8.4) *is full-rank factorization* $W = F(AF)^* \in \mathbb{C}_s^{n\times m}$ *then the following is valid:*

$$A^{(2)}_{\mathscr{R}(Q_1),\mathscr{N}(R_1P^*)} = Q_1(R_1P^*AQ_1)^{-1}R_1P^* \in A\{2,3\}_s. \tag{3.8.7}$$

Proof. Firstly we present a formal verification part (a).

It is clear that $X = Q_1(R_1P^*AQ_1)^{-1}R_1P^* \in A\{2\}_s$. In order to verify $X \in A\{4\}$ we start from

$$Q_1(R_1P^*AQ_1)^{-1}R_1P^* = Q_1Q_1^{\dagger}(R_1P^*A)^{\dagger}R_1P^*. \tag{3.8.8}$$

Since $Q_1(R_1P^*) = (GA)^*G$ we have

$$\mathscr{R}(Q_1) = \mathscr{R}((GA)^*) = \mathscr{R}\big((GA)^\dagger\big). \tag{3.8.9}$$

Later, using

$$\mathscr{R}(Q_1Q_1^\dagger) = \mathscr{R}(Q_1)$$

together with fact that $Q_1Q_1^\dagger$ is idempotent matrix, taking into account (3.8.9), immediately follows (see, for example, Lemma 1.3.1 from Ref. [80])

$$Q_1Q_1^\dagger(GA)^\dagger = (GA)^\dagger. \tag{3.8.10}$$

Since, as it is said above, matrix Q satisfies $Q^*Q = I$ we have

$$\begin{bmatrix} Q_1^* \\ Q_2^* \end{bmatrix} \begin{bmatrix} Q_1 & Q_2 \end{bmatrix} = \begin{bmatrix} Q_1^*Q_1 & Q_1^*Q_2 \\ Q_2^*Q_1 & Q_2^*Q_2 \end{bmatrix} = I$$

and so, $Q_1^*Q_2 = Q_2^*Q_1 = O$, $Q_1^*Q_1 = I_s$, $Q_2^*Q_2 = I_{n-s}$.

We now verify identity

$$(R_1P^*A)^\dagger R_1P^* = (GA)^\dagger G. \tag{3.8.11}$$

Indeed, using $Q_1^*Q_1 = I$, we obtain

$$\begin{aligned}
(R_1P^*A)^\dagger R_1P^* &= (Q_1R_1P^*A)^\dagger Q_1R_1P^* \\
&= ((GA)^*GA)^\dagger (GA)^*G \\
&= (GA)^\dagger \left(GA(GA)^\dagger\right)^* G \\
&= (GA)^\dagger GA(GA)^\dagger G \\
&= (GA)^\dagger G.
\end{aligned}$$

This implies

$$\begin{aligned}
X &= Q_1Q_1^\dagger(R_1P^*A)^\dagger R_1P^* \\
&= Q_1Q_1^\dagger(GA)^\dagger G
\end{aligned}$$

Finally, from (3.8.10) follows that

$$X = (GA)^{\dagger}G \in A\{2,4\}_s.$$

Part (b) can be verified as in the following. Using reverse order low for Moore–Penrose's inverse, we obtain

$$\begin{aligned} X &= Q_1(R_1P^*AQ_1)^{-1}R_1P^* \\ &= Q_1(AQ_1)^{\dagger}(R_1P^*)^{\dagger}R_1P^*. \end{aligned}$$

Now, applying

$$\begin{aligned} (R_1P^*)^{\dagger}R_1P^* &= (Q_1R_1P^*)^{\dagger}Q_1R_1P^* \\ &= (F(AF)^*)^{\dagger}F(AF)^* \\ &= ((AF)^*)^{\dagger}F^{\dagger}F(AF)^* \\ &= \left(AF(AF)^{\dagger}\right)^* \\ &= AF(AF)^{\dagger} \end{aligned}$$

the following representation for X can be obtained:

$$X = Q_1(AQ_1)^{\dagger}AF(AF)^{\dagger}.$$

Now, problem is to verify

$$Q_1(AQ_1)^{\dagger}AF(AF)^{\dagger} = F(AF)^{\dagger}.$$

Since $Q_1(R_1P^*) = F(AF)^*$ we have

$$\mathscr{R}(Q_1) = \mathscr{R}(F) = \mathscr{R}\left(F(AF)^{\dagger}\right).$$

Thus, $F(AF)^{\dagger} = Q_1Y$, for appropriately chosen matrix Y. In accordance with Exercise 1.3(2) from Ref. [80], we have

$$\begin{aligned} Q_1(AQ_1)^{\dagger}AF(AF)^{\dagger} &= Q_1(AQ_1)^{\dagger}AQ_1Y \\ &= Q_1Y = F(AF)^{\dagger}, \end{aligned}$$

which implies $X = F(AF)^{\dagger} \in A\{2,3\}_s$. $\square$

Complete factorization rank $W = FG$ which satisfy $F^*F = I$ called orthogonal [47]. Thus, the "full–rank" representation (3.8.4) is orthogonal.

In the following statements is shown that orthogonal "full–rank" representation based on the QR decomposition of the matrix W provides the same $A^{(2)}_{T,S}$ and the inverse representation of Lemma 3.7.1.

Corollary 3.8.7. [60] *Some prerequisites are met Lemma 3.8.4 and let $W = FG$ an arbitrary factorization full-rank matrix W. Then it is satisfied*

$$Q_1(R_1P^*AQ_1)^{-1}R_1P^* = F(GAF)^{-1}G \tag{3.8.12}$$

Proof. On the basis of the uniqueness of generalized inverse of a given core and images, applying assertion (b) Lemma 3.8.4 and Lemma 3.7.1, statement (2), we conclude

$$\begin{aligned} Q_1(R_1P^*AQ_1)^{-1}R_1P^* &= A^{(2)}_{\mathscr{R}(W),\mathscr{N}(W)} \\ &= A^{(2)}_{\mathscr{R}(F),\mathscr{N}(G)} \\ &= F(GAF)^{-1}G, \end{aligned}$$

thus the proof is complete. □

In the next lemma from Ref. [60] is shown that the external representation of the inverse $A^{(2)}_{\mathscr{R}(W),\mathscr{N}(W)}$, defined in Lemma 3.7.1,invariant to the choice of the full-rank of the matrix factorization W. In this sense, the obtained generalization consequences 3.8.7.

Lemma 3.8.8. [60] *Some prerequisites of Lemma 3.8.4 are satisfied. If $W = F_1G_1$ i $W = F_2G_2$ two different complete factorization rank matrix W, then it is valid*

$$\begin{aligned} F_1(G_1AF_1)^{-1}G_1 &= F_2(G_2AF_2)^{-1}G_2 \\ &= A^{(2)}_{\mathscr{R}(W),\mathscr{N}(W)}. \end{aligned}$$

Proof. On the basis of Theorem 2 from Ref. [47], there is invertible $s \times s$ matrices D such that

$$F_2 = F_1 D, \;\; G_2 = D^{-1} G_1.$$

This relationship implies the following:

$$F_2(G_2 A F_2)^{-1} G_2 = F_1 D \left(D^{-1} G_1 A F_1 D\right)^{-1} D^{-1} G_1.$$

As matrices D i $G_1 A F_1$ are invertible, applies the law to reverse the order $\left(D^{-1} G_1 A F_1 D\right)^{-1}$, such that

$$F_2(G_2 A F_2)^{-1} G_2 = F_1 D D^{-1} (G_1 A F_1)^{-1} D D^{-1} G_1 = F_1 (G_1 A F_1)^{-1} G_1.$$

It follows from this evidence introduced by the statements. □

Finally we found a correlation between general representation given in the Ref. [53] and representations introduced in Lemma 3.8.4.

Corollary 3.8.9. [60] *Some conditions of Lemma 3.8.4 are satisfied. Let $W = FG$ an arbitrary representation of the full-rank of the matrix W which satisfies the conditions Lemma 3.7.1. If* (3.8.4) *complete factorization rank of a matrix W, satisfied the statements are as follows:*

$$F = Q_1 D, \;\; G = D^{-1} R_1 P^*, \tag{3.8.13}$$

whereby the matrix is equal to D. $D = Q_1^* F$.

Example 3.8.2. *Consider the following matrix*

$$A = \begin{bmatrix} 1 & 2 & 3 & 4 & 1 \\ 1 & 3 & 4 & 6 & 2 \\ 2 & 3 & 4 & 5 & 3 \\ 3 & 4 & 5 & 6 & 4 \\ 4 & 5 & 6 & 7 & 6 \\ 6 & 6 & 7 & 7 & 8 \end{bmatrix} \tag{3.8.14}$$

and choose matrices

$$F=\begin{bmatrix} 3 & -2 \\ -1 & 1 \\ 0 & 0 \\ 0 & 0 \\ 0 & 0 \end{bmatrix}, \quad G=\begin{bmatrix} 1 & 0 & 0 & 0 & 0 & 0 \\ 0 & 1 & 0 & 0 & 0 & 0 \end{bmatrix}. \tag{3.8.15}$$

Suppose that the matrix W is defined as $W = FG$. *Matrices* Q_1, R_1 *and* P *that define non-trivial part of the QR decomposition* (3.8.1) *matrices* WP *are given as*

$$\{Q_1,R_1,P\} = \left\{ \begin{bmatrix} \frac{3}{\sqrt{10}} & \frac{1}{\sqrt{10}} \\ -\frac{1}{\sqrt{10}} & \frac{3}{\sqrt{10}} \\ 0 & 0 \\ 0 & 0 \\ 0 & 0 \end{bmatrix}, \begin{bmatrix} \sqrt{10} & -\frac{7}{\sqrt{10}} & 0 & 0 & 0 & 0 \\ 0 & \frac{1}{\sqrt{10}} & 0 & 0 & 0 & 0 \end{bmatrix}, I_6 \right\}.$$

Representations given in statements (b) *and* (c) *of Lemma 3.8.4 give the following* $\{2\}$*–inverse A:*

$$A^{(2)}_{\mathscr{R}(W),\mathscr{N}(W)} = \begin{bmatrix} 3 & -2 & 0 & 0 & 0 & 0 \\ -1 & 1 & 0 & 0 & 0 & 0 \\ 0 & 0 & 0 & 0 & 0 & 0 \\ 0 & 0 & 0 & 0 & 0 & 0 \\ 0 & 0 & 0 & 0 & 0 & 0 \end{bmatrix}.$$

Although $W = FG$ *and* $W = Q_1(R_1P^*)$ *are two different representations of the full-rank of the matrix W, based on the consequences of 3.8.7*

$$A^{(2)}_{\mathscr{R}(W),\mathscr{N}(W)} = Q_1(R_1P^*AQ_1)^{-1}R_1P^* = F(GAF)^{-1}G.$$

We denote now with I(n) algorithm complexity to invert given $n \times n$ matrix (as in Ref. [10]). In addition to the $\mathscr{A}(n)$ complexity mark the addition/subtraction of two $n \times n$ matrices with a $M(m,n,k)$ complex multiplications $m \times n$ matrix $n \times k$ matrix. A simpler notation $\mathcal{M}(n)$ (taken from Ref. [10]) is used instead of the $\mathcal{M}(n,n,n)$.

Remark 3.8.1. *The representation of data in claiming (d) of Lemma 3.8.4 includes $M(s,n,m)$ operations for the formation of the product matrix Q_1^*W. On the other hand, the representation defined assertion (b) Lemma 3.8.4 requires $M(s,m,m)$ operations for the formation of a dot-matrix product R_1P^*. Accordingly, the representation (d) is better than the representation (b) in the case when $n < m$. In the opposite case $m < n$ is a representation of (b) suitable in relation to the representation of (d).*

Efficient methods to calculate $A^{(2)}_{\mathcal{R}(W),\mathcal{N}(W)}$ is defined as follows. For the case $m < n$, taking into account the complexity of matrix multiplication, you'd better set of equations solved

$$R_1P^*AQ_1X = R_1P^*. \tag{3.8.16}$$

For the case $n < m$, effectively solved the following set of equations

$$Q_1^*WAQ_1X = Q_1^*W. \tag{3.8.17}$$

When you generate a matrix X, it is necessary to calculate the product matrix

$$A^{(2)}_{\mathcal{R}(Q_1),\mathcal{N}(R_1P^*)} = A^{(2)}_{\mathcal{R}(Q_1),\mathcal{N}(Q_1^*W)} = Q_1X. \tag{3.8.18}$$

Corollary 3.8.10. [60] *Let $A \in \mathbb{C}_r^{m\times n}$ be specified matrix, $s \leq r$ it is given integer and let the matrix F, G chosen as in State 3.8.3.*

(a) *If* (3.8.4) *is full-rank factorization of matrix $W = (GA)^*G \in \mathbb{C}_s^{n\times m}$ then the following is valid:*

$$A^{(2,4)}_{\mathcal{R}(Q_1),\mathcal{N}(R_1P^*)} = Q_1(R_1P^*AQ_1)^{-1}R_1P^* = (GA)^{\dagger}G \in A\{2,4\}_s. \tag{3.8.19}$$

(b) *If* (3.8.4) *is full-rank factorization of matrix* $W = F(AF)^* \in \mathbb{C}_s^{n\times m}$ *then the following is valid:*

$$A^{(2,3)}_{\mathscr{R}(Q_1),\mathscr{N}(R_1P^*)} = Q_1(R_1P^*AQ_1)^{-1}R_1P^* = F(AF)^{\dagger} \in A\{2,3\}_s. \quad (3.8.20)$$

Proof. (a) In this case, based on consequence 3.8.7 and state 3.8.3 we have

$$Q_1(R_1P^*AQ_1)^{-1}R_1P^* = (GA)^*(GA(GA)^*)^{-1}G = (GA)^{\dagger}G.$$

(b) This piece of evidence is checked in a similar way. □

Example 3.8.3. *Let be given the matrix A defined in* (3.8.14) *and choose the matrices F, G defined in* (3.8.15).

(*i*) *Full-rank representation* (3.8.4) *of matrix* $W = (GA)^*G$ *is given with*

$$\{Q_1,R_1,P\} = \left\{ \begin{bmatrix} \frac{1}{\sqrt{31}} & -2\sqrt{\frac{7}{93}} \\ \frac{2}{\sqrt{31}} & \sqrt{\frac{3}{217}} \\ \frac{3}{\sqrt{31}} & -\frac{11}{\sqrt{651}} \\ \frac{4}{\sqrt{31}} & 2\sqrt{\frac{3}{217}} \\ \frac{1}{\sqrt{31}} & \frac{17}{\sqrt{651}} \end{bmatrix}, \begin{bmatrix} \sqrt{31} & \frac{45}{\sqrt{31}} & 0 & 0 & 0 & 0 \\ 0 & \sqrt{\frac{21}{31}} & 0 & 0 & 0 & 0 \end{bmatrix}, I_6 \right\}.$$

Representation (b) from Lemma 3.8.4 gives the following $\{2,4\}$*-inverse of A:*

$$A^{(2,4)}_{\mathscr{R}(Q_1),\mathscr{N}(R_1P^*)} = Q_1(R_1P^*AQ_1)^{-1}R_1P^* = \begin{bmatrix} 1 & -\frac{2}{3} & 0 & 0 & 0 & 0 \\ -\frac{1}{7} & \frac{1}{7} & 0 & 0 & 0 & 0 \\ \frac{6}{7} & -\frac{11}{21} & 0 & 0 & 0 & 0 \\ -\frac{2}{7} & \frac{2}{7} & 0 & 0 & 0 & 0 \\ -\frac{8}{7} & \frac{17}{21} & 0 & 0 & 0 & 0 \end{bmatrix}$$

$$= (GA)^{\dagger}G$$

$$= A^{(2,4)}_{\mathscr{R}((GA)^*),\mathscr{N}(G)}.$$

(ii) Let choose the matrix W as $W = F(AF)^*$. full-rank representation (3.8.4) *of matrix W is given with*

$$\{Q_1, R_1, P\}\left\{\begin{bmatrix} \frac{3}{\sqrt{10}} & \frac{1}{\sqrt{10}} \\ -\frac{1}{\sqrt{10}} & \frac{3}{\sqrt{10}} \\ 0 & 0 \\ 0 & 0 \\ 0 & 0 \end{bmatrix}, y, I_6\right\}.$$

Representation from Lemma 3.8.4, the statements (b) leads to the following $\{2,3\}$*-inverse matrix A:*

$$A^{(2,3)}_{\mathscr{R}(Q_1),\mathscr{N}(R_1P^*)} = Q_1(R_1P^*AQ_1)^{-1}R_1P^* =$$

$$\begin{bmatrix} -\frac{59}{392} & -\frac{69}{196} & -\frac{39}{392} & -\frac{19}{392} & \frac{1}{392} & \frac{15}{49} \\ \frac{55}{392} & \frac{61}{196} & \frac{43}{392} & \frac{31}{392} & \frac{19}{392} & -\frac{9}{49} \\ 0 & 0 & 0 & 0 & 0 & 0 \\ 0 & 0 & 0 & 0 & 0 & 0 \\ 0 & 0 & 0 & 0 & 0 & 0 \end{bmatrix}$$

$$= F(AF)^\dagger$$

$$= A_{\mathscr{R}(F),\mathscr{N}((AF)^*)}.$$

$$Here, y = \begin{bmatrix} \sqrt{10} & -\frac{7}{\sqrt{10}} & \frac{37}{\sqrt{10}} & 32\sqrt{\frac{2}{5}} & \frac{91}{\sqrt{10}} & 81\sqrt{\frac{2}{5}} \\ 0 & \frac{1}{\sqrt{10}} & -\frac{1}{\sqrt{10}} & -\sqrt{\frac{2}{5}} & -\frac{3}{\sqrt{10}} & -3\sqrt{\frac{2}{5}} \end{bmatrix}$$

The following is an algorithm for calculating $A^{(2)}_{T,S}$ inverse of given matrix A.

Algorithm 3.8 Calculating $A^{(2)}_{T,S}$ inverse matrix A using QR decomposition of the matrix W. **(algorithm GenInvQR2)**

Input: The matrix A dimensions $m \times n$ rank and r.

1: Select an arbitrary fixed $n \times m$ matrix W rank $s\ leqr$.

Table 3.9a. The Algorithm Complexity of Calculating External Inverse $X = Q_1(R_1P^*AQ_1)^{-1}R_1P^*$

Term	Complexity
$\Lambda_1 = R_1P^*$	$\mathcal{M}(s,m,m)$
$\Lambda_2 = \Lambda_1 A$	$\mathcal{M}(s,m,m) + \mathcal{M}(s,m,n)$
$\Lambda_3 = \Lambda_2 Q_1$	$\mathcal{M}(s,m,m) + \mathcal{M}(s,m,n) + \mathcal{M}(s,n,s)$
$\Lambda_4 = (\Lambda_3)^{-1}$	$\mathcal{M}(s,m,m) + \mathcal{M}(s,m,n) + \mathcal{M}(s,n,s) + \mathrm{I}(s)$
$\Lambda_5 = Q_1\Lambda_4$	$\mathcal{M}(s,m,m) + \mathcal{M}(s,m,n) + \mathcal{M}(s,n,s) + \mathrm{I}(s) + \mathcal{M}(n,s,s)$
$X = \Lambda_5\Lambda_1$	$\mathcal{M}(s,m,m) + \mathcal{M}(s,m,n) + \mathcal{M}(s,n,s) + \mathrm{I}(s) + \mathcal{M}(n,s,s) + \mathcal{M}(n,s,m)$

2: Calculate QR decomposition of the matrix W form (3.8.1).

3: Determine the decomposition of the full-rank of the matrix W as in (3.8.4).

4: Solve matrix equation of the form (3.8.16) for the case $m < n$, or matrix equation (3.8.17) for the case $n < m$.

5: Determine output solution $A^{(2)}_{\mathscr{R}(Q_1),\mathscr{N}(R_1P^*)} = A^{(2)}_{\mathscr{R}(Q_1),\mathscr{N}(Q_1^*W)}$ as in (3.8.18).

3.8.1 The Complexity and Implementation of Algorithms and GenInvQR2

Algorithm *GenInvQR*2 we use for QR decomposition of matrix W of the form (3.8.4) and then calculate $A^{(2)}_{T,S}$ inverse based on statement (b) in Lemma 3.8.4. Thus, the complexity of its calculation is equal to

$$E(GenInvQR2(W)) = \mathcal{M}(s,m,m) + \mathcal{M}(s,m,n) + \mathcal{M}(s,n,s) + \mathrm{I}(s) + \mathcal{M}(n,s,s) + \mathcal{M}(n,s,m). \tag{3.8.21}$$

Table 3.9a describes the algorithm complexity of calculating external inverse, $X = Q_1(R_1P^*AQ_1)^{-1}R_1P^*$.

It is well-known that the same weight matrix inversion and matrix multiplication. More specifically, plain inverse real nonsingular $n \times n$ matrix can be calculated in time $\mathrm{I}(n) = \mathcal{O}(\mathcal{M}(n))$ [10]. Notation $\mathcal{O}(f(n))$ is described in Ref. [10], for example. In the implementation we use the usual methods for matrix multiplication, so that $\mathcal{M}(m,n,k) = m \cdot n \cdot k$. This implies, on the basis of (3.8.21)

$$E(GenInvQR2(W)) = 2s^2n + sm(m+2n) + \mathrm{I}(s) \approx 2s^2n + sm(m+2n) + s^3. \tag{3.8.22}$$

For the special case $W = A^*$, algorithm *GenInvQR2* takes *QR* decomposition of matrix A^* of form (3.8.4) and calculates $A^\dagger$. On the other hand, the algorithm *qrginv* begins with *QR* decomposition of the matrix A forms

$$A = Q_A R_A P_A^*$$

and calculate Moore–Penrose's inverse

$$A^\dagger = P_A R_A^\dagger Q_A^*.$$

Its complexity calculation is given by

$$E(qrginv) = pinv(r,n) + \mathcal{M}(n,n,r) + \mathcal{M}(n,r,m) = rn(m+n) + pinv(r,n), \tag{3.8.23}$$

where $pinv(r,n)$ indicates the complexity of calculating the rectangular $r \times n$ of real matrix. For the special case $W = A^*$ of algorithm *GenInvQR2*, its complexity is

$$E(GenInvQR2(A^*)) = r^2(m+2n) + 2rmn + pinv(r,r). \tag{3.8.24}$$

Implementation Algorithm *GenInvQR2* in `Matlab` is surrounded by the herein in its entirety. The system matrix equation (3.8.16) solved using `Matlab` 'backslash' operator. All results are generated for the case $m < n$. Similar results can be obtained for the case $m > n$ under a mild modification of the above `Matlab` code.

```
function ATS2 =ATS2(A)
U=A';
[Q,R,P] = qr(U);
r = rank(U);
Q1 = Q(:,1:r);
R1 = R(1:r,:);
G = R1*P';
X = (G*A*Q1)\(G);
ATS2 = Q1*X;
```

The index of the given matrix can be calculated using the following `Matlab` functions:

```
function [k,r,A1] = index(A)
k = 0;
n = length(A);
r= n;
A0 = A;
r1 = rank(A);
A1 = zeros(n);
while r~=r1
r=r1;
A1 = A;
A = A * A0;
r1 = rank(A);
k = k+1;
end
```

3.8.2 From GenInvQR2 to qrginv and reverse

Suppose that $AP = QR$ is given QRdecomposition, defining the matrix Q with orthonormal columns, $R = \begin{bmatrix} R_1 \\ 0 \end{bmatrix}$, where R_1 is full-rank row (see [82], Theorem 3.3.11). Then $AP = Q_1R_1$, after the trimming of rows and columns, and taking into account the ranking of the A. Based on Ref. [33]

(comment after Theorem 4) is satisfied

$$\begin{aligned} A^\dagger &= PR^\dagger Q^* = P \begin{bmatrix} R_1^\dagger & 0 \end{bmatrix} Q^* \\ &= PR_1^\dagger Q_1^*. \end{aligned} \tag{3.8.25}$$

The Ref. [60] is defined by two representations Moore–Penrose's inverse. The first representation is defined as a separate case $W = A^*$ GenInvQR2 algorithm. More specifically, it ranks from QR decomposition $A^* = Q_1(R_1P^*)$ of matrix A^* and then calculates the Moore–Penrose's work on the basis of the inverse (b) of Lemma 3.8.4.

Another representation of Moore–Penrose's inverse is based on the decomposition of the matrix A^*, obtained from the QR decomposition of the matrix A. If $AP = Q_1R_1$, $Q_1^*Q_1 = I$ then $A^* = (PR_1^*)Q_1^*$ is full-rank factorization A^*. From Lemma 3.8.4 we get

$$GenInvQR2(A^*) = PR_1^* \,(Q_1^*APR_1^*)^{-1}\, Q_1^*. \tag{3.8.26}$$

It is not hard to check that representation (3.8.26) produces the same result as the representation (3.8.25). Really, applying

$$\begin{aligned} GenInvQR2(A^*) &= PR_1^* \,(Q_1^*APR_1^*)^{-1}\, Q_1^* \\ &= PR_1^* \,(Q_1^*Q_1R_1P^*PR_1^*)^{-1}\, Q_1^* \\ &= PR_1^*(R_1R_1^*)^{-1}Q_1^* \end{aligned}$$

and taking into account that the matrix R_1 the full-rank of species, it follows that $GenInvQR2(A^*) = PR_1^\dagger Q_1^*$, which is exactly equal to (3.8.25).

3.8.3 Numerical Experiments for the Moore–Penrose's Inverse

In this section, we mention the results of using the language of `Matlab` calculations, as well as for testing the accuracy of the results obtained. All numerical tasks were carried out using `Matlab` R2009 and the environment on Intel(R) Pentium(R) Dual CPU T2310 1.46 GHz 1.47 GHz 32-bit system with 2 GB RAM memory on Windows Vista Home Premium Operating system.

In Ref. [84], authors Wei-Wu specified somewhat iterative method, after testing with different sets matrix and concluded that the most accurate equation is (3.8) in Ref. [84], from the standpoint of accuracy and execution time. Therefore, only this method from Ref. [84] is included in the numerical tests.

Testing algorithms *GenInvQR*2 *qrginv* and Wei-Wu method is executed separately for random singular matrix and the singular test matrix with a large number of the conditioning of *Matrix Computation Toolbox* (see [24]). We also tested the proposed methods in the rows of the matrix from Matrix Market collection (see [39]), and thereby obtained very efficiently accurate results in all cases.

Finally, numerical results are given for the case of inverse Drazin's, i.e., in case of $W = A^k$, where k index matrix A. Note that the sign '-' signifies a long processing time executive tonal algorithm. Tolerance for iterative methods (3.8) from Ref. [83] is set at 10^{-6}. Implementation of *GenInvQR*2 method `Matlab`, and `Matlab` function to calculate matrix index are listed in the appendix. Results for accuracy are determined by applying the following four relationships that characterize Moore–Penrose's inverse:

$$AA^{\dagger}A = A, \quad A^{\dagger}AA^{\dagger} = A^{\dagger}, \quad (AA^{\dagger})^* = AA^{\dagger}, \quad (A^{\dagger}A)^* = A^{\dagger}A.$$

Randomly Generated Singular Test Matrices

We compared the performance of the proposed method *GenInvQR*2 the algorithm *qrginv* from Ref. [33] method the above Eq. (3.8) in Ref. [83] on a set of random singular matrix ranking 2^n, $n = 8, 9, 10, 11$.

In Table 3.9, the accuracy of the method has been tested on the basis of error 2-norm standard. As before, the *rank* (resp. *Cond*) column contains the rank (resp. Conditional number) of each tested matrices. It is evident that the proposed methods *GenInvQR*2 gives an excellent approximation in each test performed. Developed method performs fast and accurate calculations, and therefore can be a good alternative for the calculation of Moore–Penrose's inverse.

Singular Test Matrix from Matrix Computation Toolbox

Fitness numbers of these matrices are of the order which is in the rank of 10^{15} and 10^{135}. For the purpose of comparative results, we also applied the introduced *GenInvQR*2 algorithm, *qrginv* function and Wei-Wu method. Given that this matrix of relatively small size, to measure the time required for each algorithm to accurately calculate Moore–Penrose's inverse, each the algorithm is running 100 times. Specified time is the arithmetic mean of the time obtained in the 100 repetitions. Errors are listed in Tables 3.9–3.14.

It is interesting to observe the complementarity algorithms *GenInvQR*2 and *qrginv* in Tables 3.13–3.14. More precisely, results from the application of the algorithm are *GenInvQR*2 $\{1,2,3\}$ inverses while *Qrginv* algorithm gives results that are $\{1,2,4\}$ inverses. The algorithm uses the *GenInvQR*2 *QR* decomposition matrices A^* with algorithm starts *qrginv* *QR* Decomposition of the matrix A. This fact may account for the differences in the convergence of the algorithm *GenInvQR*2 for the case $W = A^*$ and the algorithm *qrginv*. If using equation (3.8.26), then no complementary results in the Tables 3.13–3.14.

Table 3.9b. Error Time Calculations and Random Singular Matrix

rank\kond.	Metod	Vreme	$\|AA^\dagger A - A\|_2$	$\|A^\dagger AA^\dagger - A^\dagger\|_2$	$\|AA^\dagger - (AA^\dagger)^*\|_2$	$\|A^\dagger A - (A^\dagger A)^*\|_2$
2^8\1.14e+04	GenInvQR2	0.15	1.04 e-12	3.2 e-12	2.7 e-12	2.1 e-12
	qrginv	0.15	2.4 e-12	6.2 e-12	4.33 e-12	5.22 e-13
	Wei,Wu	5.21	4.21 e-12	4.234 e-08	2.34 e-13	1.02 e-11
2^9\1.29e+04	GenInvQR2	1.18	6.35 e-12	7.02 e-11	5.33 e-12	1.11 e-11
	qrginv	1.17	2.22 e-12	1.82 e-12	2.98 e-12	4.22 e-13
	Wei,Wu	44.33	5.02 e-11	1.213 e-07	2.532 e-13	6.632 e-12
2^10\6.31e+05	GenInvQR2	9.48	6.01 e-11	1. e-09	1.22 e-10	8.71 e-11
	qrginv	8.399	1.15 e-10	3.166 e-10	1.05 e-10	1.504 e-11
	Wei,Wu	432.12	2.40 e-10	3.196 e-10	7.19 e-12	6.01 e-09
2^11\8.88e+04	GenInvQR2	76.13	5.209 e-11	2.103 e-10	5.2148 e-11	8.16 e-11
	qrginv	61.75	2.93 e-11	8.51 e-12	2.093 e-11	2.243 e-12
	Wei,Wu	-				

Table 3.9c. Error Time Calculation for `chow` (rank=199, Cond=8.01849 e+135)

Metod	Vreme	$\|AA^{\dagger}A-A\|_2$	$\|A^{\dagger}AA^{\dagger}-A^{\dagger}\|_2$	$\|AA^{\dagger}-(AA^{\dagger})^{*}\|_2$	$\|A^{\dagger}A-(A^{\dagger}A)^{*}\|_2$	
GenInvQR2	0.024	9.117 e-13	2.0 e-13	1.662 e-13	1.262 e-12	6.057 e-13
qrginv		1.114 e-13		8.249 e-15		1.5847 e-14
Wei,Wu	0.039	4.437 e-13	1.08 e-13	1.419	3.456 e-13	5.501 e-13

Table 3.10. Error Time Calculation for `cycol` (rank=50, Cond=2.05 e+48)

Method	Timing	$\|AA^{\dagger}A-A\|_2$	$\|A^{\dagger}AA^{\dagger}-A^{\dagger}\|_2$	$\|AA^{\dagger}-(AA^{\dagger})^{*}\|_2$	$\|A^{\dagger}A-(A^{\dagger}A)^{*}\|_2$
GenInvQR2	0.014	3.203 e-14	8.110 e-17	1.552 e-15	3.205 e-15
qrginv	0.0179	3.22 e-14	1.189 e-15	6.095 e-17	1.258 e-15
Wei,Wu	0.401	5.218 e-07	1.91 e-09	2.39 e-15	0.891 e-15

Table 3.11. Error Time Calculation for `gearmat` (rank=199, Cond=3.504 e+17)

Method	Timing	$\|AA^{\dagger}A-A\|_2$	$\|A^{\dagger}AA^{\dagger}-A^{\dagger}\|_2$	$\|AA^{\dagger}-(AA^{\dagger})^{*}\|_2$	$\|A^{\dagger}A-(A^{\dagger}A)^{*}\|_2$
GenInvQR2	0.0142	4.031 e-15	4.037 e-13	0.94 e-14	1.060 e-13
qrginv	1.26	8.853 e-14	2.01 e-14	4.56 e-15	5.18 e-15
Wei,Wu	0.0233	3.168 e-15	1.425 e-13	4.15 e-13	4.28 e-10

Table 3.12. Error Time Calculation for `kahan` (rank=199, Cond=2.30018 e+24)

Method	Timing	$\|AA^{\dagger}A-A\|_2$	$\|A^{\dagger}AA^{\dagger}-A^{\dagger}\|_2$	$\|AA^{\dagger}-(AA^{\dagger})^*\|_2$	$\|A^{\dagger}A-(A^{\dagger}A)^*\|_2$
GenInvQR2	0.011	3.063 e-14	1.035 e-09	6.237 e-10	3.13 e-14
qrginv	0.035	2.00 e-05	1.666 e-09	0.523	7.174 e-15
Wei,Wu	3.87	6.64 e-15	2.52 e-10	3.449 e-10	3.142 e-14

Table 3.13. Error Time Calculation for `lotkin` (rank=19, Cond=8.97733 e+21)

Method	Timing	$\|AA^{\dagger}A-A\|_2$	$\|A^{\dagger}AA^{\dagger}-A^{\dagger}\|_2$	$\|AA^{\dagger}-(AA^{\dagger})^*\|_2$	$\|A^{\dagger}A-(A^{\dagger}A)^*\|_2$
GenInvQR2	0.061	8.45 e-06	9.34 e-08	3.03 e-11	0.2384
qrginv	0.03	8.02 e-06	8.21 e-09	0.023	1.007 e-11
Wei,Wu	-				

Table 3.14. Error Time Calculation for `prolate` (rank=117, Cond=5.61627 e+17)

Method	Timing	$\|AA^{\dagger}A-A\|_2$	$\|A^{\dagger}AA^{\dagger}-A^{\dagger}\|_2$	$\|AA^{\dagger}-(AA^{\dagger})^*\|_2$	$\|A^{\dagger}A-(A^{\dagger}A)^*\|_2$
GenInvQR2	0.151	1.137 e-06	1.134 e-06	5.557 e-11	0.0377
qrginv	0.114	1.16 e-06	2.7480 e-07	0.027	4.19 e-11
Wei,Wu	-				

3.8.4 Numerical Experiments for Drazin Inverse

In Ref. [60] the numerical experiments are given for the calculation of Drazin inverse A^D square matrix A, using the algorithm for determining $A^{(2)}_{T,S}$ inverse. In the case of inverse Drazin, we take that $W = A^k$, where k index of matrix A. To check the accuracy of the results, we used the following three relations from the definition of Drazin inverse:

$$A^D A = AA^D, \quad A^D AA^D = A^D, \quad A^{k+1}A^D = A^k.$$

First, give the example of the calculation Drazin inverse square matrix of small dimensions, using methods *GenInvQR*2.

Example 3.8.4. Let be given singular matrix

$$A = \begin{bmatrix} 2 & 0 & 0 \\ 0 & 1 & 1 \\ 0 & -1 & -1 \end{bmatrix}.$$

Then, index of matrix A is equal to $\text{ind}(A) = 2$ and

$$W = A^2 = \begin{bmatrix} 4 & 0 & 0 \\ 0 & 0 & 0 \\ 0 & 0 & 0 \end{bmatrix}.$$

QR decomposition of the full-rank of the matrix W is determined by the

$$W = Q_1 R_1 P^* = \begin{bmatrix} 1 \\ 0 \\ 0 \end{bmatrix} \begin{bmatrix} 4 & 0 & 0 \end{bmatrix} \begin{bmatrix} 1 & 0 & 0 \\ 0 & 1 & 0 \\ 0 & 0 & 1 \end{bmatrix}.$$

Accordingly, based on the work under (b) Lemma 3.8.4 consequences 3.8.5, we have

$$A^D = Q_1(R_1 P^* A Q_1)^{-1} R_1 P^* = \begin{bmatrix} \frac{1}{2} & 0 & 0 \\ 0 & 0 & 0 \\ 0 & 0 & 0 \end{bmatrix}$$

Tables 3.15 and 3.16 compares the errors and time calculations the above method with the method presented in Ref. [83]. Re-use set of eight test Matrix dimensions 200×200 received function `matrix` Matrix Computation Toolbox (Table 3.15). We also used a series random singular matrix of rank 2^n, $n = 8, 9, 10, 11$ (Table 3.16).

As is evident from Table 3.15, iterative methods Wei-Wu ([83]) converges in most cases. Therefore, he does not give exact numerical answer. In contrast, $A_T^{S(2)}$ the algorithm gives very accurate results in most cases and is very effective.

Wei-Wu iterative method does not work for a reasonable time for the case of randomly generated singular matrix, so that this method excluded from the comparisons listed in Table 3.15.

3.9 Symbolic Computations of $A_{T,S}^{(2)}$ Inverses Using *QDR* Factorization

The main motivation of work [61] is the representation of Moore–Penrose's inverse $A^\dagger$ from Ref. [33]. Methods introduced in Ref. [33] is derived from the *QR* decomposition of the matrix A. The Ref. [61] obtained two extensions of this algorithm. The first generalization is that the algorithm is applicable for the calculation of sire class $A_{T,S}^{(2)}$ inverses, not only for computing Moore–Penrose inverse. Instead of the *QR* decomposition of the matrix A, *QDR* decomposition of an arbitrarily selected matrix W is used, and thus obtained full-rank factorization. Selecting *QDR* decomposition is critical in eliminating the appearance of the square root of the elements of the *QR* decomposition.

Computations which include square roots entries are inappropriate for symbolic and algebraic computations. Symbolic implementation expres-

Table 3.15. Error Time Calculations and Random Singular Matrices

Index	rank	Timing	$\|AA^D-A^DA\|_2$	$\|A^DAA^D-A^D\|_2$	$\|A^{k+1}A^D-A^k\|_2$
1	2^8	0.624	1.71 e-10	2.339 e-09	3.4369 e-10
1	2^9	2.2436	5.235 e-11	1.251 e-10	2.111 e-10
1	2^{10}	12.6	2.166 e-09	1.388 e-08	8.013 e-09
1	2^{11}	175.01	2.1031 e-09	7.143 e-09	8.04 e-09

sions $\sqrt{\sum_{i=0}^{q} A_i s^i}$ which include constant matrices A_i, $i = 0,\ldots,q$ and unknown s is a very complicated problem in procedural programming languages and a job whose execution requires a lot processor time in packages for symbolic computations. In doing so, the square root of some polynomial matrix often occur in generating QR factorization. Generating expressions containing square roots can be avoided using QDR decomposition. This technique is of essential importance in calculating symbolic polynomial. Similarly as in the Ref. [58], motivation in the Ref. [61] was to take advantage of free-roots decomposition in symbolic calculations and broaden algorithm GenInvQR2 on a set of polynomial matrices. The question is: which is the main reason for the replacement of LDL^* with the QR decomposition. The main drawback LDL^* decomposition is that it is applicable only to symmetric positive definite matrix. This sets a limit of obstacles result from Ref. [58] on the set $\{1,2,3\},\{1,2,4\}$-inverses and Moore–Penrose's inverse. Representations listed in the work [61] are applicable for a broader set of external inverse defined core and images. For these reasons, instead of LDL^* decomposition, we will use QDR factorization of rational matrices to avoid receiving elements with roots. Evidently, the resulting algorithm is very suitable for implementation in procedural programming languages, for free-roots elements of the matrix and the basic simplification method which involves calculating GCDs two polynomials.

Basic *QDR* matrix factorization A generates three matrix: matrix Q of equal rank ranks from A, diagonal matrix D and R matrix in several stages. Here is the proposed algorithm which we introduced in Ref. [61] a direct calculation of *QDR* decomposition of full-rank, where the matrix Q formed no zero column, R is generated no zero rows, a diagonal matrix D has no zero rows and zero columns. *QDR* decomposition products an additional diagonal matrix compared to *QR* decomposition, but returns as a result of the matrix with elements without roots, better suited for still symbolic calculation.

Algorithm 3.9 Full-rank *QDR* factorization of rational matrix A

Input: Rational matrix $A \in \mathbb{C}(x)_s^{n\times m}$.

1: Construct three zero matrices: $Q \in \mathbb{C}(x)^{n\times s}$, $D \in \mathbb{C}(x)^{s\times s}$, $R \in \mathbb{C}(x)^{s\times m}$.

2: For $i = \overline{1,s}$ repeat

  2.1: $B := A - QDR$.

  2.2: Determine first the next non-zero column matrix B and notate it by c.

  2.3: Set i-th column of matrix Q to the column c.

  2.4: For $j = \overline{i,m}$ set $R_{i,j}$ to inner vector product c and j-th column of B.

  2.5: Set element $D_{i,i}$ to inverse of square of 2-norm of the column c.

Note that the matrix equation $A = QDR + B$ solved at each step of the algorithm, where it starts at $A = B$. In the end, $B = 0$ and $A = QDR$. It is also the matrix R upper triangular, a column of the matrix Q contain orthogonal base area column matrix A. The implementation of this algorithm in `MATHEMATICA` is listed in the Appendix.

In many cases, Gram-Schmidt's algorithm with pivoting column is essential. In every step of columns of the matrix B with the highest 2-norm is selected instead of selecting the first non-zero column. Then the matrix R is upper triangular matrix with permuted columns, while the columns of Q again contain again contain orthogonal base space of columns of A.

Example 3.9.1. Observe these two matrices:

$$F = \begin{bmatrix} 3 & -2 \\ -1 & 1 \\ 0 & 0 \\ 0 & 0 \\ 0 & 0 \end{bmatrix}, \quad G = \begin{bmatrix} 1 & 0 & 0 & 0 & 0 & 0 \\ 0 & 1 & 0 & 0 & 0 & 0 \end{bmatrix}.$$

and matrix W as

$$W = FG = \begin{bmatrix} 3 & -2 & 0 & 0 & 0 & 0 \\ -1 & 1 & 0 & 0 & 0 & 0 \\ 0 & 0 & 0 & 0 & 0 & 0 \\ 0 & 0 & 0 & 0 & 0 & 0 \\ 0 & 0 & 0 & 0 & 0 & 0 \end{bmatrix}.$$

QDR decomposition of W is determined by

$$\begin{bmatrix} 3 & -2 & 0 & 0 & 0 & 0 \\ -1 & 1 & 0 & 0 & 0 & 0 \\ 0 & 0 & 0 & 0 & 0 & 0 \\ 0 & 0 & 0 & 0 & 0 & 0 \\ 0 & 0 & 0 & 0 & 0 & 0 \end{bmatrix} =$$

$$\begin{bmatrix} 3 & \frac{1}{10} \\ -1 & \frac{3}{10} \\ 0 & 0 \\ 0 & 0 \\ 0 & 0 \end{bmatrix} \begin{bmatrix} \frac{1}{10} & 0 \\ 0 & 10 \end{bmatrix} \begin{bmatrix} 10 & -7 & 0 & 0 & 0 & 0 \\ 0 & \frac{1}{10} & 0 & 0 & 0 & 0 \end{bmatrix}.$$

Example 3.9.2. Consider the following matrix

$$W = \begin{bmatrix} -3-4x^2 & 2-7x & 4 \\ -9x & -3+3x^2 & -5 \\ -2x+9x^2 & 9x^2 & -5 \end{bmatrix}.$$

By applying the Algorithm 3.9 we get the following matrix from *QDR* decomposition matrix W:

$$Q = \begin{bmatrix} -3-4x^2 & \frac{x\left(81+170x-694x^2+657x^3-747x^4+324x^5\right)}{9+109x^2-36x^3+97x^4} & \frac{3x\left(90-217x+53x^2+3742x^3-1388x^4-3444x^5+5565x^6+2052x^7\right)}{81+324x-704x^2-2956x^3+8998x^4-11880x^5+13824x^6-486x^7+2169x^8} \\ -9x & -3+3x^2+\frac{9x\left(-6+48x-8x^2-17x^3+81x^4\right)}{9+109x^2-36x^3+97x^4} & -\frac{x\left(-180+2279x+3163x^2-10909x^3+8706x^4+10329x^5-14904x^6+8208x^7\right)}{81+324x-704x^2-2956x^3+8998x^4-11880x^5+13824x^6-486x^7+2169x^8} \\ x(-2+9x) & \frac{x\left(-12+231x-448x^2+1019x^3-9x^4+144x^5\right)}{9+109x^2-36x^3+97x^4} & \frac{3\left(-135-552x+675x^2+2603x^3-2674x^4-1292x^5+4108x^6-60x^7+912x^8\right)}{81+324x-704x^2-2956x^3+8998x^4-11880x^5+13824x^6-486x^7+2169x^8} \end{bmatrix},$$

$$D = \begin{bmatrix} \frac{1}{9+109x^2-36x^3+97x^4} & 0 & 0 \\ 0 & \frac{9+109x^2-36x^3+97x^4}{81+324x-704x^2-2956x^3+8998x^4-11880x^5+13824x^6-486x^7+2169x^8} & 0 \\ 0 & 0 & \frac{81+324x-704x^2-2956x^3+8998x^4-11880x^5+13824x^6-486x^7+2169x^8}{\left(45+94x-113x^2-15x^3+228x^4\right)^2} \end{bmatrix},$$

$$R = \begin{bmatrix} 9+109x^2-36x^3+97x^4 & \dfrac{-6+48x-8x^2-17x^3+81x^4}{\frac{81+324x-704x^2-2956x^3+8998x^4-11880x^5+13824x^6-486x^7+2169x^8}{9+109x^2-36x^3+97x^4}} \\ 0 & 0 \\ 0 & -12+55x-61x^2 - \dfrac{\frac{-135-654x+1135x^2+716x^3+1882x^4+6048x^5+879x^6}{9+109x^2-36x^3+97x^4}}{\frac{\left(45+94x-113x^2-15x^3+228x^4\right)^2}{81+324x-704x^2-2956x^3+8998x^4-11880x^5+13824x^6-486x^7+2169x^8}} \end{bmatrix}.$$

Many famous representation of various generalized inverses defined rank and for generalized inverses defined core and images. The most useful representation for the research is the next "full-rank" external representation of the inverse pre-specified core and pictures from Ref. [53].

In the special case representation full-rank Drazin's inverse A^D, based on an arbitrary decomposition of the full-rank A^l, $l \geq \text{ind}(A)$,is introduced into the Ref. [67].

Alternative explicit expression for the inverse General $A^{(2)}_{T,S}$, which is based on the use of group inverse, was introduced in operation [84]. Characterization theorems about the representation and the term limits for $A^{(2)}_{T,S}$ obtained in Ref. [84] by the application of this representation.

Authors of Ref. [7] performed basic representation theorem and generalized representation of outer inverse $A^{(2)}_{T,S}$. Based on this representation, somewhat SPECIFIC representation and iterative methods for calculating the $A^{(2)}_{T,S}$ is presented in Ref. [7].

The following assertion determines the full-rank of external representation of the inverse of the given rank, zero space, and rank, the same general form as in Section 3.7.1. Assertion is valid for rational matrix and is based on a complete factorization rank matrix W resulting from the decomposition of QDR defined in Algorithm 3.9.

Lemma 3.9.1. *Let $A \in \mathbb{C}(x)_r^{m\times n}$ be given matrix. For an arbitrary matrix $W \in \mathbb{C}(x)_s^{n\times m}$, $s \leq r$, consider its QDR decomposition got with Algorithm 3.9, of the form*

$$W = QDR,$$

where $Q \in \mathbb{C}(x)_s^{n\times s}$, $D \in \mathbb{C}(x)_s^{s\times s}$ is a diagonal matrix, and $R \in \mathbb{C}(x)_s^{s\times m}$ is an upper triangular matrix. Let the happy condition

$$rank(W) = rank(RAQ) = s. \tag{3.9.1}$$

define a set of

$$\mathbb{C}_s(W) = \{x_c \in \mathbb{C} |\ rank(W) = \text{rank}(W(x_c)) = \text{rank}(R(x_c)A(x_c)Q(x_c)) = s\}. \tag{3.9.2}$$

Then the following assertion satisfied on the set $\mathbb{C}_s(W)$:

$$\begin{aligned} A^{(2)}_{\mathscr{R}(Q),\mathscr{N}(R)} &= Q(RAQ)^{-1}R \\ &= A^{(2)}_{\mathscr{R}(W),\mathscr{N}(W)}. \end{aligned} \tag{3.9.3}$$

Proof. Obviously, factorization

$$W = QDR = (QD)(R), \tag{3.9.4}$$

represents a complete factorization rank matrix W at the set $\mathbb{C}_s(W)$. Given that D and RAQ invertible matrix, they meet a property reverse order, i.e., $(RAQD)^{-1} = D^{-1}(RAQ)^{-1}$. Now, the first identity in (3.9.3) follows from state 3.7.1 and identity

$$\begin{aligned} QD(RAQD)^{-1}R &= Q(RAQ)^{-1}R \\ &= A^{(2)}_{\mathscr{R}(Q),\mathscr{N}(R)}. \end{aligned}$$

It is evident now

$$A^{(2)}_{\mathscr{R}(W),\mathscr{N}(W)} = A^{(2)}_{\mathscr{R}(Q),\mathscr{N}(R)}$$

satisfied on the set $\mathbb{C}_s(W)$ from (3.9.4). □

Remark 3.9.1. Note that for a given matrix $A \in \mathbb{C}(x)_r^{m\times n}$ arbitrarily selected matrix $W \in \mathbb{C}(x)_s^{n\times m}$, $s \leq r$, products corresponding outer core and the inverse of a given image forms (3.9.3), where (3.9.4) QDR decomposition of the matrix W. External inverse $A^{(2)}_{\mathscr{R}(Q),\mathscr{N}(R)}$ is a function of the set $\mathbb{C}(x)$. External inverse elements, marked with g_{ij}, are also features on the set $\mathbb{C}(x)$. Then the domain functions $A^{(2)}_{\mathscr{R}(Q),\mathscr{N}(R)}$ designated as $C_s(W)\bigcap_{i,j}\mathrm{Dom}(g_{ij})$, where $\mathrm{Dom}(g_{ij})$ means the domain functions g_{ij}.

Taking into account the representation (1.1.1)–(1.1.3) the main foreign inverses, we get the following representation.

Corollary 3.9.2. *For a given matrix* $A \in \mathbb{C}(x)_r^{m\times n}$ *and a selected matrix* $W \in \mathbb{C}(x)_s^{n\times m}$ *with QDR decomposition defined in* (3.9.4). *Next the statements are valid at the meeting* $\mathbb{C}_s(W)\bigcap_{i,j}\mathrm{Dom}(g_{ij})$:

$$\begin{aligned}
&(a)\quad A^{(2)}_{\mathscr{R}(Q),\mathscr{N}(R)} = A^{\dagger} \ \textit{za slucaj}\ W = A^{*};\\
&(b)\quad A^{(2)}_{\mathscr{R}(Q),\mathscr{N}(R)} = A^{\dagger}_{M,N} \ \textit{za slucaj}\ W = A^{\sharp};\\
&(c)\quad A^{(2)}_{\mathscr{R}(Q),\mathscr{N}(R)} = A^{\#} \ \textit{za slucaj}\ W = A;\\
&(d)\quad A^{(2)}_{\mathscr{R}(Q),\mathscr{N}(R)} = A^{D} \ \textit{za slucaj}\ W = A^{k},\ k \geq \mathrm{ind}(A);\\
&(e)\quad A^{(2)}_{\mathscr{R}(Q),\mathscr{N}(R)} = A^{(-1)}_{(L)} \ \textit{za slucaj}\ \mathscr{R}(W) = L,\ \mathscr{N}(W) = L^{\perp};\\
&(f)\quad A^{(2)}_{\mathscr{R}(Q),\mathscr{N}(R)} = A^{(\dagger)}_{(L)} \ \textit{za slucaj}\ \mathscr{R}(W) = S,\ \mathscr{N}(W) = S^{\perp}.
\end{aligned} \tag{3.9.5}$$

Numerical very stable approach for calculating (3.9.3) solving the set of equations

$$RAQX = R \tag{3.9.6}$$

and then calculate the product matrix

$$A^{(2)}_{\mathscr{R}(Q),\mathscr{N}(R)} = QX. \tag{3.9.7}$$

Algorithm 3.10 Calculation of the inverse matrix $A^{(2)}_{T,S}$ using QDR faktorizacije (**Algorithm GenInvQDR2**)

Input: Rational matrix $A(x) \in \mathbb{C}(x)_r^{m\times n}$. Select an arbitrary fixed $n \times m$ matrix W normal rank $s \leq r$.

1: Generate QDR decomposition of the full-rank of the matrix W by applying Algorithm 3.9. Transform rational matrix Q, R in the general form of (3.8.1).

2: Transform rational matrix $M = RAQ$ in form:

$$M = \frac{1}{p(x)} M_1,$$

where $p(x)$ is corresponding polynomial and M_1 is polynomial matrix.

3: Find inverse of matrix M_1 using algorithm 3.2 from Ref. [74]. Generalize inverse of matrix $N = M^{-1} = p(x)M_1^{-1}$, and transform in the following form:

$$N_{ij}(x) = \frac{\sum\limits_{k=0}^{\overline{n}_q} \overline{n}_{k,i,j} x^k}{\sum\limits_{k=0}^{\overline{\overline{n}}_q} \overline{\overline{n}}_{k,i,j} x^k}.$$

4: Introduce substitute $\overline{\alpha}_q = \overline{q}_q + \overline{n}_q + \overline{r}_q$, $\overline{\overline{\alpha}}_q = \overline{\overline{q}}_q + \overline{\overline{n}}_q + \overline{\overline{r}}_q$, and for $i = \overline{1,n}$, $j = \overline{1,m}$ repeat Steps 5.1 – 5.5.

5.1: For $k = \overline{1, \min\{j,s\}}$, $l = \overline{1,s}$ perform the following calculations

$$\begin{aligned}
\overline{\alpha}_{t,i,j,k,l} &= \sum_{t_2=0}^{t_1} \sum_{t_3=0}^{t_1-t_2} \overline{q}_{t_2,i,l} \overline{n}_{t_1-t_2-t_3,l,k} \overline{r}_{t_3,k,j}, \quad 0 \leq t \leq \overline{\alpha}_q, \\
\overline{\overline{\alpha}}_{t,i,j,k,l} &= \sum_{t_2=0}^{t_1} \sum_{t_3=0}^{t_1-t_2} \overline{\overline{q}}_{t_2,i,l} \overline{\overline{n}}_{t_1-t_2-t_3,l,k} \overline{\overline{r}}_{t_3,k,j}, \quad 0 \leq t \leq \overline{\overline{\alpha}}_q.
\end{aligned}$$

5.2: Polynomial of denominator (i,j)-th of element $A^{(2)}_{\mathscr{R}(Q),\mathscr{N}(R)}$ as

$$\text{LCM}\left\{ \sum_{t=0}^{\overline{\overline{\alpha}}_q} \overline{\overline{\alpha}}_{t,i,j,k,l} x^t \,\middle|\, k = \overline{1, \min\{j,s\}},\ l = \overline{1,s} \right\}, \tag{3.9.8}$$

and denote with $\overline{\overline{\Gamma}}_{ij}(x) = \sum_{t=0}^{\overline{\overline{\Gamma}}_q} \overline{\overline{\gamma}}_{t,i,j} x^t$.

5.3: For each $k = \overline{1, \min\{j,s\}}$, $l = \overline{1,s}$ calculate polynomial $\overline{\overline{\Gamma}}_{ij}(x) / \sum_{t=0}^{\overline{\overline{\alpha}}_q} \overline{\overline{\alpha}}_{t,i,j,k,l} x^t$ and denote with $\Gamma_{i,j,k,l}(x) = \sum_{t=0}^{\overline{\overline{\Gamma}}_q - \overline{\overline{\alpha}}_q} \gamma_{t,i,j,k,l} x^t$.

5.4: Calculate polynomial of numerator (i,j)-th element of inverse $A^{(2)}_{\mathscr{R}(Q),\mathscr{N}(R)}$ on the following way

$$\overline{\Gamma}_{ij}(x) = \sum_{t=0}^{\overline{\overline{\Gamma}}_q - \overline{\overline{\alpha}}_q + \overline{\alpha}_q} \left(\sum_{k=1}^{\min\{j,s\}} \sum_{l=1}^{s} \sum_{t_1=0}^{t} \overline{\alpha}_{t_1,i,j,k,l} \gamma_{t-t_1,i,j,k,l} \right) x^t,$$

5.5: Set (i,j)-th element of matrix $A^{(2)}_{\mathscr{R}(Q),\mathscr{N}(R)}$ on values $\overline{\Gamma}_{ij}(x) / \overline{\overline{\Gamma}}_{ij}(x)$.

Polynomial $P(x)$ is primitive if all its coefficients are mutually co-prime. Rational functions can be stored as ordered pairs primitive numerators and denominators. Notice that simplification is crucial in Steps 3 and 5.5, where quotients two polynomials are evaluated. Function *Simplify*[ ] performs a sequence algebraic and other transformations on a given expression and returns simplest form it finds [86]. Package `MATHEMATICA` is appropriate for symbolic computations and has built-functions for manipulating with unevaluated expressions.

In procedural programming languages, this simplification can be done by using greatest common divisor two polynomials. Fast *gcd* algorithm considering Chinese remainder theorem and simple Euclid algorithm can be used for finding greatest common divisor two polynomials. Coefficients in intermediate results can expand greatly in performing Euclid algorithm for polynomial reminder. But, notice that one can evaluate primitive part remainder. However, primitive part computations requires many greatest common divisors coefficients which can also be large. Thus, Chinese Remainder Algorithm (CRA) can be used for reconstruction *gcd* coefficients back to integers.

3.9.1 Experiments with Polynomial and Rational Matrices

In the next few examples will illustrate the algorithm and then test some implementations like comparing time execution on different test matrices.

Example 3.9.3. Consider the following polynomial matrix

$$A=\begin{bmatrix} -4x^2-3 & 2-7x & 4 \\ -9x & 3x^2-3 & -5 \\ 9x^2-2x & 9x^2 & -5 \\ -4x^2-3 & 2-7x & 4 \end{bmatrix},$$

$$W=\begin{bmatrix} 3 & 7x & 4 & 5 \\ -9x & 3x^2-3 & 5 & x+5 \\ -6 & -14x & -8 & -10 \end{bmatrix}.$$

Matrix W was chosen quite arbitrarily, but with appropriate dimensions. We have that $r = rank(A) = 3$, $s = rank(W) = 2$. Algorithm 3.9 gives the following QDR factorization of W:

$$Q=\begin{bmatrix} 3 & \frac{9x\left(8x^2-1\right)}{9x^2+5} \\ -9x & \frac{15\left(8x^2-1\right)}{9x^2+5} \\ -6 & -\frac{18x\left(8x^2-1\right)}{9x^2+5} \end{bmatrix}, \quad D=\begin{bmatrix} \frac{1}{81x^2+45} & 0 \\ 0 & \frac{9x^2+5}{45(1-8x^2)^2} \end{bmatrix},$$

$$R=\begin{bmatrix} 9\left(9x^2+5\right) & 3x\left(44-9x^2\right) & 60-45x & 75-9x(x+5) \\ 0 & \frac{45\left(1-8x^2\right)^2}{9x^2+5} & \frac{15(12x+5)\left(8x^2-1\right)}{9x^2+5} & \frac{15(16x+5)\left(8x^2-1\right)}{9x^2+5} \end{bmatrix}.$$

Expression `X = QDRAlgorithm[A,W]` leads to the next of outer inverse matrix A:

$$X = \begin{bmatrix} -\frac{3(72x^4+108x^3-148x^2-3x+19)}{636x^6+777x^5+9129x^4-9265x^3-198x^2+749x+352} & -\frac{3(72x^4+108x^3-148x^2-3x+19)}{636x^6+777x^5+9129x^4-9265x^3-198x^2+749x+352} & \frac{108x^4-875x^3+297x^2+98x-48}{636x^6+777x^5+9129x^4-9265x^3-198x^2+749x+352} & \frac{108x^4-875x^3+297x^2+98x-48}{636x^6+777x^5+9129x^4-9265x^3-198x^2+749x+352} \\ -\frac{3(172x^3-241x^2+39x+35)}{636x^6+777x^5+9129x^4-9265x^3-198x^2+749x+352} & -\frac{3(172x^3-241x^2+39x+35)}{636x^6+777x^5+9129x^4-9265x^3-198x^2+749x+352} & \frac{212x^4+199x^3+702x^2-59x-144}{636x^6+777x^5+9129x^4-9265x^3-198x^2+749x+352} & \frac{212x^4+199x^3+702x^2-59x-144}{636x^6+777x^5+9129x^4-9265x^3-198x^2+749x+352} \\ \frac{6(72x^4+108x^3-148x^2-3x+19)}{636x^6+777x^5+9129x^4-9265x^3-198x^2+749x+352} & \frac{6(72x^4+108x^3-148x^2-3x+19)}{636x^6+777x^5+9129x^4-9265x^3-198x^2+749x+352} & -\frac{2(108x^4-875x^3+297x^2+98x-48)}{636x^6+777x^5+9129x^4-9265x^3-198x^2+749x+352} & -\frac{2(108x^4-875x^3+297x^2+98x-48)}{636x^6+777x^5+9129x^4-9265x^3-198x^2+749x+352} \end{bmatrix}.$$

Based on Lemma 3.9.1 we get

$$X = A^{(2)}_{\mathscr{R}(Q),\mathscr{N}(R)} = A^{(2)}_{\mathscr{R}(W),\mathscr{N}(W)}.$$

Using `MATHEMATICA` function `NullSpace` we get

$$\mathscr{N}(R) = \begin{bmatrix} -\frac{8x^2-35x-15}{9(8x^2-1)} & -\frac{16x+5}{3(8x^2-1)} & 0 & 1 \\ -\frac{12x^2-35x-12}{9(8x^2-1)} & -\frac{12x+5}{3(8x^2-1)} & 1 & 0 \end{bmatrix}.$$

Also, just to check that =

$$\mathscr{R}(Q) = \left\{ \frac{9x\left(8x^2-1\right)z}{9x^2+5} + 3y, \frac{15\left(8x^2-1\right)z}{9x^2+5} - 9xy, -\frac{18x\left(8x^2-1\right)z}{9x^2+5} - 6y \right\},$$

whereby y, z are arbitrary complex numbers.

On the other hand, expression `QDRAlgorithm[A,Transpose[A]]` implies calculating the corresponding case $W = A^T$, and generates Moore–Penrose's inverse of matrix A:

$$A^{\dagger}=\begin{bmatrix} -\frac{15(2x^2+1)}{456x^4-30x^3-226x^2+188x+90} & \frac{-36x^2+35x-10}{228x^4-15x^3-113x^2+94x+45} & \frac{12x^2-35x-2}{228x^4-15x^3-113x^2+94x+45} & -\frac{15(2x^2+1)}{456x^4-30x^3-226x^2+188x+90} \\ \frac{5x(9x+7)}{456x^4-30x^3-226x^2+188x+90} & \frac{16x^2-8x-15}{228x^4-15x^3-113x^2+94x+45} & \frac{20x^2+36x+15}{228x^4-15x^3-113x^2+94x+45} & \frac{5x(9x+7)}{456x^4-30x^3-226x^2+188x+90} \\ \frac{3x(9x^3+25x^2-9x+2)}{456x^4-30x^3-226x^2+188x+90} & -\frac{x(36x^3-63x^2+59x-4)}{228x^4-15x^3-113x^2+94x+45} & \frac{3(4x^4+20x^2-6x-3)}{228x^4-15x^3-113x^2+94x+45} & \frac{3x(9x^3+25x^2-9x+2)}{456x^4-30x^3-226x^2+188x+90} \end{bmatrix}.$$

The same result is obtained and for $W = A$ and $W = A^k$, $k \geq 2$, given that Drazin and group inverse equal to $A^{\dagger}$.

In the following example we will observe a rational matrix of which the Moore–Penrose's inverse and group inverse different.

Example 3.9.4. Next rational matrix

$$A_1=\begin{bmatrix} \frac{144(-1+8x^2)}{7(-36-25x+164x^2)} & 0 & 0 & \frac{108+175x+4x^2}{252+175x-1148x^2} \\ \frac{75+372x}{252+175x-1148x^2} & 1 & 0 & \frac{3(25+124x)}{7(-36-25x+164x^2)} \\ -\frac{99(-1+8x^2)}{7(-36-25x+164x^2)} & 0 & 1 & \frac{99(-1+8x^2)}{7(-36-25x+164x^2)} \\ \frac{144(-1+8x^2)}{7(-36-25x+164x^2)} & 0 & 0 & \frac{108+175x+4x^2}{252+175x-1148x^2} \end{bmatrix}$$

is equal to the product AX_1 of matrices A from Example 3.9.3 and its external inverse X_1:

$$X_1=\begin{bmatrix} -\frac{6(-452+395x+3418x^2-6852x^3+7344x^4)}{7(-1620-4509x+9098x^2+18781x^3-26365x^4-8160x^5+37392x^6)} & \frac{-10+35x-36x^2}{45+94x-113x^2-15x^3+228x^4} & \frac{-2-35x+12x^2}{45+94x-113x^2-15x^3+228x^4} & \frac{3(356+1665x+3616x^2-11954x^3+3208x^4)}{7(-1620-4509x+9098x^2+18781x^3-26365x^4-8160x^5+37392x^6)} \\ \frac{6(435+784x-2434x^2+976x^3+6000x^4)}{7(-1620-4509x+9098x^2+18781x^3-26365x^4-8160x^5+37392x^6)} & \frac{-15-8x+16x^2}{45+94x-113x^2-15x^3+228x^4} & \frac{15+36x+20x^2}{45+94x-113x^2-15x^3+228x^4} & \frac{-2610-13524x-2861x^2+26449x^3+15660x^4}{7(-1620-4509x+9098x^2+18781x^3-26365x^4-8160x^5+37392x^6)} \\ \frac{3(-297-982x+6631x^2+9197x^3-34020x^4+33264x^5+7200x^6)}{7(-1620-4509x+9098x^2+18781x^3-26365x^4-8160x^5+37392x^6)} & \frac{x(4-59x+63x^2-36x^3)}{45+94x-113x^2-15x^3+228x^4} & \frac{3(-3-6x+20x^2+4x^4)}{45+94x-113x^2-15x^3+228x^4} & \frac{3(297+478x-4713x^2-11626x^3+17045x^4-6139x^5+3132x^6)}{7(-1620-4509x+9098x^2+18781x^3-26365x^4-8160x^5+37392x^6)} \end{bmatrix}.$$

Matrix A_1 is idempotent, and obviously $\text{ind}(A_1) = 1$. Moore–Penrose's inverse A_1 is generated in case $W = A_1^T$ is equal to the next

$$A_1^\dagger = \begin{bmatrix} \frac{9(2873+4500x-26336x^2-11200x^3+108320x^4)}{47826+93600x-318719x^2+1400x^3+1954384x^4} & \frac{8100+53301x+65400x^2+1488x^3}{47826+93600x-318719x^2+1400x^3+1954384x^4} & \frac{99(-108-175x+860x^2+1400x^3+32x^4)}{47826+93600x-318719x^2+1400x^3+1954384x^4} & \frac{9(2873+4500x-26336x^2-11200x^3+108320x^4)}{47826+93600x-318719x^2+1400x^3+1954384x^4} \\ \frac{3(-900-89x+50600x^2+143344x^3)}{95652+187200x-637438x^2+2800x^3+3908768x^4} & \frac{42201+37800x-457103x^2+1400x^3+1954384x^4}{47826+93600x-318719x^2+1400x^3+1954384x^4} & -\frac{297(-25-124x+200x^2+992x^3)}{47826+93600x-318719x^2+1400x^3+1954384x^4} & \frac{3(-900-89x+50600x^2+143344x^3)}{95652+187200x-637438x^2+2800x^3+3908768x^4} \\ \frac{99(36-175x-1444x^2+1400x^3+9248x^4)}{95652+187200x-637438x^2+2800x^3+3908768x^4} & -\frac{297(-25-124x+200x^2+992x^3)}{47826+93600x-318719x^2+1400x^3+1954384x^4} & \frac{38025+93600x-161903x^2+1400x^3+1327120x^4}{47826+93600x-318719x^2+1400x^3+1954384x^4} & \frac{99(36-175x-1444x^2+1400x^3+9248x^4)}{95652+187200x-637438x^2+2800x^3+3908768x^4} \\ \frac{42642+118800x-110783x^2-200200x^3+622672x^4}{95652+187200x-637438x^2+2800x^3+3908768x^4} & \frac{432(-25-124x+200x^2+992x^3)}{47826+93600x-318719x^2+1400x^3+1954384x^4} & \frac{14256(1-8x^2)^2}{47826+93600x-318719x^2+1400x^3+1954384x^4} & \frac{42642+118800x-110783x^2-200200x^3+622672x^4}{95652+187200x-637438x^2+2800x^3+3908768x^4} \end{bmatrix}.$$

For $W = A_1$, get a group that is equal to the inverse of the initial matrix A_1.

Example 3.9.5. Let the data matrix

$$A = \begin{bmatrix} x-1 & x-1 & 2x-2 \\ x & x & x \end{bmatrix}$$

and choose matrix $W = A^*$. Then it is valid that $rank(W) = 2$ i $C_2(W) = \mathbb{C} \setminus \{1, 0\}$, since

$$\text{rank}(W(1)) = \text{rank}(W(0)) = 1 < 2.$$

Notation $W(1)$ and $W(0)$ indicate a constant matrix obtained by replacing the symbol x values $x = 1$ i $x = 0$, respectively. Moore–Penrose's inverse of matrix A is given with

$$A^\dagger = \begin{bmatrix} \frac{1}{2-2x} & \frac{1}{x} \\ \frac{1}{2-2x} & \frac{1}{x} \\ \frac{1}{x-1} & -\frac{1}{x} \end{bmatrix}.$$

Obviously $A^\dagger$ is not defined in case $x \in \{1, 0\}$ (or equivalent, in case $x \notin C_2(W)$).

3.9.2 Implementation Details and Results of the Testing

Complexity of *QDR* decomposition is equal to complexity of *QR* decomposition. Implementation of Algorithm *GenInvQDR*2 was made in the package MATHEMATICA. Created two main functions, called QDRDecomposition [A_List] and QDRAlgorithm[A_List], testing and checking.

Normal rank the set of rational matrix is determined by standard MATHEMATICA function MatrixRank working with numerical and symbolic matrices with Ref. [86]. Algorithm 3.9 implemented with the following MATHEMATICA function.

```
QDRDecomposition[A_List] :=
 Module[{i, j, n = Length[A[[1]]], m = Length[A], Q, c = 0, Diag, R, B, A1},
  Q = Table[0, {n}, {m}]; Diag = R = Table[0, {n}, {n}]; B = A;
  For[i = 1, i ≤ n, i++,
   A1 = Transpose[B];
   While[(i ≤ n) && (Norm[A1[[i]]] == 0), c++; i++];
   c = 0;
   If[i > n, Break[]];
   NextCol = A1[[i]];
   Q[[i]] = Simplify[A1[[i]]];
   For[j = i, j ≤ n, j++,
    R[[i]][[j]] = Simplify[NextCol.A1[[j]]];];
   Diag[[i]][[i]] = Simplify[1 / (Norm[NextCol]^2), Element[x, Reals]];
   B = A - Transpose[Q].Diag.R;
  ];
  Q = Drop[Q, -c]; Diag = Drop[Diag, -c];
  Diag = Drop[Transpose[Diag], -c]; R = Drop[R, -c];
  Return[{ExpandNumerator[Together[Transpose[Q]]],
    ExpandNumerator[ExpandDenominator[Diag]], ExpandNumerator[R]}];
 ];
```

Implementation of Algorithm *GenInvQDR* the data following a free code, whereby the matrix equation (3.9.6) solves with standard MATHEMATICA function Inverse:

```
GenInvQDR[A_List, W_List] :=
Module[{N1, Q, Diag, R},
{Q, Diag, R} = QDRDecomposition[W];
N1 = Inverse[R.A.Q] // Simplify;
```

```
Return[Simplify[Q.N1.R]];
];
```

Example 3.9.6. Comparison of different algorithms for symbolic calculation Moore–Penrose's inverse is specified in Table 3.16. CPU time is taken as a criterion for the comparison of those algorithms. Algorithm *GenInvQDR*2 has been tested for a special case of $W = A^*$. The processing times are obtained by using `MATHEMATICA` implementation of different algorithms to calculate the pseudoinverse in some test matrices from Ref. [88].

The first row of the table contains the name of the test matrix from Ref. [88], generated by Zielke, with three groups of test matrix (A, S and F) taken into consideration. The last type includes the time obtained *GenInvQDR*2 algorithm. Note that *GenInvQDR*2 algorithm less efficient than partitioning methods or Leverier-Faddeev algorithm, due to several matrix multiplication, where inter-results and coefficients can significantly grow. Compared to algorithms based on Cholesky and LDL* decomposition, the algorithm is superior to us.

3.10 Singular Value-Decomposition

Penrose has proposed a direct method for calculating Moore–Penrose's-inverse using singular-value decomposition.

Theorem 3.10.1. *If a values-specified singular decomposition Matrix A*

$$A = U^T \begin{bmatrix} \Sigma & \mathbb{O} \\ \mathbb{O} & \mathbb{O} \end{bmatrix} V = U^T \begin{bmatrix} \sigma_1 & & & \\ & \ddots & & \mathbb{O} \\ & & \sigma_r & \\ & \mathbb{O} & & \mathbb{O} \end{bmatrix} V,$$

Table 3.16. Average Time of Calculation (in seconds) for a Few Algorithms and Algorithm 3.10

Test matrix [88]	A_5	A_6	A_7	S_5	S_6	S_7	F_5	F_6	F_7
PseudoInverse [86]	0.23	0.55	1.01	0.1	0.45	0.79	0.45	0.69	1.14
Partitioning [65]	0.1	0.12	0.45	0.1	0.22	0.44	0.1	0.23	0.63
Lev.-Faddeev [29, 66]	0.0	0.12	0.34	0.0	0.11	0.23	0.0	0.1	0.42
LDLGInverse [58]	8.6	72.1	143.4	5.2	300.1	–	12.1	175.4	–
Cholesky [69]	8.54	79.8	233.8	5.66	343.5	–	13.2	202.1	–
GenInvQDR	1.21	13.19	197.12	0.1	1.12	71.02	1.15	14.4	231.18
GenInvQDR2	1.12	13.13	191.23	0.11	0.9	65.2	1.23	13.2	224.2

where U and V orthogonal matrix, then

$$A^{\dagger} = V^T \begin{bmatrix} \Sigma^{-1} & \mathbb{O} \\ \mathbb{O} & \mathbb{O} \end{bmatrix} U = V^T \begin{bmatrix} \frac{1}{\sigma_1} & & & \\ & \ddots & & \mathbb{O} \\ & & \frac{1}{\sigma_r} & \\ & \mathbb{O} & & \mathbb{O} \end{bmatrix} U.$$

In these equations are $\sigma_1 \leq \cdots \leq \sigma_r > 0$ singular values of the matrix A, i.e., eigenvalues of matrix A^*A.

Using the full-rank representation of generalized inverse,singular value decomposition to calculate full-rank factorization is obtained following the presentation of generalized inverse [62].

Statement 3.10.1. *Let $A \in \mathbb{C}_r^{n\times n}$ and $A = U \begin{bmatrix} \Sigma & \mathbb{O} \\ \mathbb{O} & \mathbb{O} \end{bmatrix} V^*$ singular value decomposition of the A, where U and V is unique matrix, and $\Sigma = diag(\sigma_1, \ldots \sigma_r)$ contains a singular value of A. Then,*
$X \in A\{2\} \Leftrightarrow X = VY(Z^{r|}\Sigma Y_{r|})^{-1}ZU^*, rank(Z^{r|}\Sigma Y_{r|}) = t,$

$gde\, je Y \in \mathbb{C}^{n\times t}, Z \in \mathbb{C}^{t\times m}, t \in \{1,\ldots,r\};$

In case of $Y \in \mathbb{C}^{n\times t}$ i $Z \in \mathbb{C}^{r\times m}$ we get

$X \in A\{1,2\} \Leftrightarrow X = VY(Z^{r|}\Sigma Y_{r|})^{-1}ZU^*, rank(Z^{r|}\Sigma Y_{r|}) = r;$
$X \in A\{1,2,3\} = X = VY(\Sigma Y_{r|})^{-1}(U^*)_{r|}, rank(\Sigma Y_{r|}) = r;$
$X \in A\{1,2,4\} = X = V^{r|}(Z^{r|}\Sigma)^{-1}ZU^*, rank(Z_{r|}\Sigma) = r;$
$A^{\dagger} = V^{r|}\Sigma^{-1}(U^*)_{r|} = V \begin{bmatrix} \Sigma & \mathbb{O} \\ \mathbb{O} & \mathbb{O} \end{bmatrix} U^*;$

$A^{\#}$ exists if and only if $(V^*)_{r|}U^{r|}\Sigma$ is invertible, in this case

$$A^{\#} = U^{r|}\Sigma((V^*)_{r|}U^{r|}\Sigma)^{-2}(V^*)_{r|}.$$

Singular decomposition of a values in MATHEMATICA can be implemented using functions *SingularValues.* The result of expression `SungularValues[m]` is list $\{u, w, v\}$, where w is list of non-zero singular values of m, while m matrix can be written in the form of `Transpose[u].DiagonalMatrix[w].v.`

Now the calculation of Moore–Penrose's inverse based on singular Axiological decomposition can implement the following functions:

```
SVDMP[a_]:=
    Block[{b=a,u,w,v},
    {u,w,v}=SingularValues[b];
    Return[Transpose[v].DiagonalMatrix[1/w].u];
    ]
```

Example 3.10.1.

```
In[1]:= a={{3,1,4,9},{1,2,3,4},{0,-2,-2,0},{-1,0,-1,-4.}}
In[2]:= {u,w,v}=SingularValues[a]
Out[2]= {{{-0.836633, -0.425092, 0.0921527, 0.33294},
> {0.206738, -0.480072, 0.793008, -0.312935},
> {0.507251, -0.505464, -0.171211, 0.676675}},
> {12.3346, 3.26459, 0.447011},
> {{-0.26494, -0.151697, -0.416637, -0.856276},
> {0.138786, -0.716605, -0.577819, 0.36516},
> {0.759745, -0.360739, 0.399006, -0.365308}}}
In[3]:= Transpose[u].DiagonalMatrix[w].v
Out[3]= {{3., 1., 4., 9.}, {1., 2., 3., 4.},
> {-1.47045*10^-16, -2., -2., 1.10995*10^-15 },
> {-1., 2.74832*10^-16, -1., -4.}}
In[4]:= x=Transpose[v].DiagonalMatrix[1/w].u
Out[4]= {{0.888889, -0.87037, -0.259259, 1.12963},
> {-0.444444, 0.518519, -0.037037, -0.481481},
> {0.444444, -0.351852, -0.296296, 0.648148},
```

```
> {-0.333333, 0.388889, 0.222222, -0.611111}}
```

then there is a nonsingular matrix T, such that

$$a.x.a = \begin{bmatrix} 3. & 1. & 4. & 9. \\ 1. & 2. & 3. & 4. \\ 3.33067*10^{-16} & -2 & -2 & 0. \\ -1. & -2.22045*10^{-16} & -1. & -4 \end{bmatrix},$$

$$x.a.x = \begin{bmatrix} 0.888889 & -0.87037 & -0.259259 & 1.12963 \\ -0.444444 & 0.518519 & -0.037037 & -0.481481 \\ 0.444444 & -0.351852 & -0.296296 & 0.648148 \\ 0.333333 & 0.388889 & 0.222222 & -0.611111 \end{bmatrix},$$

$$ax = \begin{bmatrix} 1. & -5.55112*10^{-17} & -4.44089*10^{-16} & 8.88178*10^{-16} \\ -2.22045*10^{-16} & 0.666667 & -0.333333 & -0.333333 \\ 4.44089*10^{-16} & -0.333333 & 0.666667 & -0.333333 \\ -2.22045*10^{-16} & -0.333333 & -0.333333 & 0.666667 \end{bmatrix},$$

$$xa = \begin{bmatrix} 0.666667 & -0.333333 & 0.333333 & 8.88178*10^{-16} \\ -0.333333 & 0.666667 & 0.333333 & -2.22045*10^{-16} \\ 0.333333 & 0.333333 & 0.666667 & 4.44089*10^{-16} \\ 0. & -2.22045*10^{-16} & -2.22045*10^{-16} & 1. \end{bmatrix},$$

3.10.1 Householder Decomposition

Pretpostavka 3.10.1.1. *If* $UA = \begin{bmatrix} G \\ \mathbb{O} \end{bmatrix}$ *Householder decomposition od* $A \in \mathbb{C}_r^{m\times n}$*,where* U *is unique and* G *is upper echelon, we get*
$X \in A\{2\} \Leftrightarrow X = Y(Z(U^*)^{r|}GY)^{-1}Z, rank(Z(U^*)^{r|}GY) = t,$
where $Y \in \mathbb{C}^{n\times t}, \quad Z \in \mathbb{C}^{t\times m}, \quad t \in \{1,\dots,r\};$
In case $Y \in \mathbb{C}^{n\times r}$ *i* $Z \in \mathbb{C}^{r\times m}$ *dobijamo*
$X \in A\{1,2\} \Leftrightarrow X = Y(Z(U^*)^{r|}GY)^{-1}Z, rank(Z(U^*)^{r|}GY) = r;$
$X \in A\{1,2,3\} \Leftrightarrow X = Y(GY)^{-1}U_{r|}, rank(GY) = r;$
$X \in A\{1,2,4\} \Leftrightarrow X = G^*(Z(U^*)^{r|}GG^*)^{-1}Z, rank(Z(U^*)^{r|}GG^*) = r;$

$A^{\dagger} = G^{*}(GG^{*})^{-1}U_{r|} = G^{\dagger}U_{r|}$
$A^{\#}$ *always exists and* $A^{\#} = (U^{*})^{r|}(GG^{*})^{-2}U_{r|}$.

3.10.2 Canonical Form of a Matrix

Next the method for calculating Drazin's encourages pseudoinverse of Campbell and Meyer. It can be used Jordan canonical form.

Theorem 3.10.2. *If* $A \in \mathbb{C}^{n\times n}$ *i* $ind(A) = k > 0$*, then there is a nonsingular matrix* T*, such that*

$$A = T \begin{bmatrix} R & \mathbb{O} \\ \mathbb{O} & N \end{bmatrix} T^{-1},$$

where R is nonsingular matrix, and N nullpotent matrix with index k. For any matrix T, R and N for which is valid (1.2.1) Drazin's Pseudoinverse of matrix A is

$$A_d = T \begin{bmatrix} R^{-1} & \mathbb{O} \\ \mathbb{O} & \mathbb{O} \end{bmatrix} T^{-1}.$$

Note that instead of the R matrix,a matrix A, obtained from nonsingular blocks Jordan canonical forms can be usedwhile the role of matrix N can be used matrix built and of nilpotent blocks Jordan canonical form of A.

Here Jordan representation based on the canonical form and rational canonical form are mentioned.

3.11 Generalized Inverses of Block Matrices

Theorem 3.11.1. *For a given matrix* $A \in \mathbb{C}_r^{m\times n}$*, regular matrices R and G, permutation matrix E and F and the unitary matrices U and V of appro-*

priate dimensions, so that is valid:

$$(T_1)\quad RAG = \begin{bmatrix} I_r & \mathbb{O} \\ \mathbb{O} & \mathbb{O} \end{bmatrix} = N_1 \quad (T_2)\quad RAG = \begin{bmatrix} B & \mathbb{O} \\ \mathbb{O} & \mathbb{O} \end{bmatrix} = N_2$$

$$(T_3)\quad RAF = \begin{bmatrix} I_r & K \\ \mathbb{O} & \mathbb{O} \end{bmatrix} = N_3 \quad (T_4)\quad EAG = \begin{bmatrix} I_r & \mathbb{O} \\ K & \mathbb{O} \end{bmatrix} = N_4$$

$$(T_5)\quad UAG = \begin{bmatrix} I_r & \mathbb{O} \\ \mathbb{O} & \mathbb{O} \end{bmatrix} = N_1 \quad (T_6)\quad RAV = \begin{bmatrix} I_r & \mathbb{O} \\ \mathbb{O} & \mathbb{O} \end{bmatrix} = N_1$$

$$(T_7)\quad UAV = \begin{bmatrix} B & \mathbb{O} \\ \mathbb{O} & \mathbb{O} \end{bmatrix} = N_2 \quad (T_8)\quad UAF = \begin{bmatrix} B & K \\ \mathbb{O} & \mathbb{O} \end{bmatrix} = N_5$$

$$(T_9)\quad EAV = \begin{bmatrix} B & \mathbb{O} \\ K & \mathbb{O} \end{bmatrix} = N_6$$

$$(T_{10a}) EAF = \begin{bmatrix} A_{11} & A_{11}T \\ SA_{11} & SA_{11}T \end{bmatrix},$$

where S and T multipliers obtained from the condition [44]

$$T = A_{11}^{-1}A_{12}, \qquad S = A_{21}A_{11}^{-1};$$

$$(T_{10b})\quad EAF = \begin{bmatrix} A_{11} & A_{12} \\ A_{21} & A_{22} \end{bmatrix} = \begin{bmatrix} A_{11} & A_{12} \\ A_{21} & A_{21}A_{11}^{-1}A_{12} \end{bmatrix} [77]$$

T_{11} *Transformation similarities square matrix* [50]

$$RAR^{-1} = RAEE^*R^{-1} = \begin{bmatrix} I_r & K \\ \mathbb{O} & \mathbb{O} \end{bmatrix} E^*R^{-1} = \begin{bmatrix} T_1 & T_2 \\ \mathbb{O} & \mathbb{O} \end{bmatrix}.$$

Regular R matrices and permutation matrix E can be obtained by commission of transformations which are analogous transformations of the types of matrices A over unit matrix I_m, while the regular matrix G and permutation matrix F can be obtained by the implementation of analog transformations implemented over the columns of the matrix A, the unit matrix I_n.

Implementation of block representation (T_3) u MATHEMATICA is given in the following procedure.

```
t3[a_]:=Block[{m,n,r,f,b=a,i=1,j,max=1,g,h,p1,p2},
  {m,n}=Dimensions[b];
  r=IdentityMatrix[m];    f=IdentityMatrix[n];
  While[max != 0,
  max=Abs[b[[i,i]]]; p1=i; p2=i;
  Do[Which[max<Abs[b[[g,h]]],
  max=Abs[b[[g,h]]];p1=g;p2=h
  ],    {g,i+1,m},{h,i+1,n}];
  If[max!=0,
  r=ChRow[r,i,p1];f=ChColumn[f,i,p2];
  b=ChRow[b,i,p1];b=ChColumn[b,i,p2];
  r=MultRow[r,i,1/b[[i,i]]];
  b=MultRow[b,i,1/b[[i,i]]];
  Do[Which[j!=i,r=AddRow[r,j,i,-b[[j,i]]];
  b=AddRow[b,j,i,-b[[j,i]]]], {j,m}];
  i++ ] ];
  {r,f}   ];MatrixQ[a]
```

Block decomposition of (T_1) can be generated by using transformations (T_3) twice:

$$R_1AF_1 = \begin{bmatrix} I_r & K \\ \mathbb{O} & \mathbb{O} \end{bmatrix} = N_3, \quad R_2N_3^TF_2 = \begin{bmatrix} I_r & \mathbb{O} \\ \mathbb{O} & \mathbb{O} \end{bmatrix} = N_1.$$

Then, regular matrix R, G can be calculated as follows:

$$N_1 = N_1^T = F_2^T N_3 R_2^T = F_2^T R_1 A F_1 R_2^T \Rightarrow R = F_2^T R_1, \quad G = F_1 R_2^T.$$

Implementation representation (T_1) thus be realized in the following way

```
t1[a_]:=
    Block[{r1,f1,r2,f2,b=a},
    {r1,f1}=t3[b];
    {r2,f2}=t3[Transpose[r1.b.f1]];
    {Transpose[f2].r1,f1.Transpose[r2]}
    ];MatrixQ[a]
```

Block decomposition of (T_4) can be implemented using transformations (T_3) the matrix A^T:

$$R_1 A^T F_1 = \begin{bmatrix} I_r & K \\ \mathbb{O} & \mathbb{O} \end{bmatrix}.$$

Thus,

$$F_1^T A R_1^T = \begin{bmatrix} I_r & \mathbb{O} \\ K_1^T & \mathbb{O} \end{bmatrix},$$

which implies $E = F_1^T, \quad G = R_1^T$.

```
t4[a_]:=
    Block[{r1,f1,b=a},
    {r1,f1}=t3[Transpose[b]];
    {Transpose[f1],Transpose[r1]}
    ];MatrixQ[a]
```

Example 3.11.1. Let the data matrix

$$A = \begin{bmatrix} 1 & 2 & 1 \\ 0 & 6 & 3 \\ -1 & 4 & 2 \end{bmatrix}$$

A series of transformations is as follows:

$$\left[\begin{array}{ccccccccc} 1 & 2 & 1 & 1 & 0 & 0 & 1 & 0 & 0 \\ 0 & 6 & 3 & 0 & 1 & 0 & 0 & 1 & 0 \\ -1 & 4 & 2 & 0 & 0 & 1 & 0 & 0 & 1 \end{array}\right] \rightarrow \left[\begin{array}{ccccccccc} 6 & 0 & 3 & 0 & 1 & 0 & 0 & 1 & 0 \\ 2 & 1 & 1 & 1 & 0 & 0 & 1 & 0 & 0 \\ 4 & -1 & 2 & 0 & 0 & 1 & 0 & 0 & 1 \end{array}\right] \rightarrow$$

$$\left[\begin{array}{ccccccccc} 1 & 0 & \frac{1}{2} & 0 & \frac{1}{6} & 0 & 0 & 1 & 0 \\ 2 & 1 & 1 & 1 & 0 & 0 & 1 & 0 & 0 \\ 4 & -1 & 2 & 0 & 0 & 1 & 0 & 0 & 1 \end{array}\right] \rightarrow \left[\begin{array}{ccccccccc} 1 & 0 & \frac{1}{2} & 0 & \frac{1}{6} & 0 & 0 & 1 & 0 \\ 0 & 1 & 0 & 1 & -\frac{1}{3} & 0 & 1 & 0 & 0 \\ 0 & -1 & 0 & 0 & -\frac{2}{3} & 1 & 0 & 0 & 1 \end{array}\right] \rightarrow$$

$$\left[\begin{array}{ccccccccc} 1 & 0 & \frac{1}{2} & 0 & \frac{1}{6} & 0 & 0 & 1 & 0 \\ 0 & 1 & 0 & 1 & -\frac{1}{3} & 0 & 1 & 0 & 0 \\ 0 & 0 & 0 & 1 & -1 & 1 & 0 & 0 & 1 \end{array}\right].$$

From the second and third matrix R i F:

$$R = \begin{bmatrix} 1 & \frac{1}{6} & 0 \\ 1 & -\frac{1}{3} & 0 \\ 1 & -1 & 1 \end{bmatrix}, \qquad F = \begin{bmatrix} 0 & 1 & 0 \\ 1 & 0 & 0 \\ 0 & 0 & 1 \end{bmatrix}.$$

Next the method for calculating matrix S, T i A_{11}^{-1} is given in Ref. [44]:

$$R = \begin{bmatrix} A_11 & A_{12} & I \\ A_{21} & A_{22} & \mathbb{O} \end{bmatrix} \rightarrow \begin{bmatrix} I_r & A_{11}^{-1}A_{12} & A_{11}^{-1} \\ \mathbb{O} & A_{22} - SA_{12} & -S \end{bmatrix} = \begin{bmatrix} I_r & T & A_{11}^{-1} \\ \mathbb{O} & \mathbb{O} & -S \end{bmatrix}.$$

Obviously $T = K$.

Example 3.11.2. For matrix

$$A = \begin{bmatrix} -1 & 0 & 1 & 2 \\ -1 & 1 & 0 & -1 \\ 0 & -1 & 1 & 3 \\ 0 & 1 & -1 & -3 \\ 1 & -1 & 0 & 1 \\ 1 & 0 & -1 & -2 \end{bmatrix}$$

we can get

$$A_{11}^{-1} = \begin{bmatrix} -1 & 0 \\ -1 & 1 \end{bmatrix}, \quad S = \begin{bmatrix} 1 & -1 \\ -1 & 1 \\ 0 & -1 \\ -1 & 0 \end{bmatrix}, \quad T = \begin{bmatrix} -1 & -2 \\ -1 & -3 \end{bmatrix}.$$

3.11.1 Block Representation of Generalized Inverse

From the decomposition of block matrices followed by a whole series of representations different class of generalized inverse. Below is an outline of block known representation for the various classes of generalized inverses, which are developed in Refs. [54, 89].

Theorem 3.11.2. *If $A \in \mathbb{C}_r^{m\times n}$, and if $R = \begin{bmatrix} R_1 \\ R_2 \end{bmatrix}$ i $G = \begin{bmatrix} G_1 & G_2 \end{bmatrix}$, matrix which A transformed into normal form $A = R^{-1}\begin{bmatrix} I_r & \mathbb{O} \\ \mathbb{O} & \mathbb{O} \end{bmatrix} G^{-1}$ then:*

$$
\begin{aligned}
A^{(1)} =& G\begin{bmatrix} I_r & X \\ Y & Z \end{bmatrix} R,\\
A^{(1,2)} =& G\begin{bmatrix} I_r & X \\ Y & YX \end{bmatrix} R = G\begin{bmatrix} I_r \\ Y \end{bmatrix}\begin{bmatrix} I_r & X \end{bmatrix} R,\\
A^{(1,3)} =& G\begin{bmatrix} I_r & -R_1R_2^{\dagger} \\ Y & Z \end{bmatrix} R,\\
A^{(1,4)} =& G\begin{bmatrix} I_r & X \\ -G_2^{\dagger}G_1 & Z \end{bmatrix} R,
\end{aligned}
$$

$$
\begin{aligned}
A^{(1,2,3)} =& G\begin{bmatrix} I_r & -R_1R_2^{\dagger} \\ Y & -YR_1R_2^{\dagger} \end{bmatrix} R = G\begin{bmatrix} I_r \\ Y \end{bmatrix}\begin{bmatrix} I_r, & -R_1R_2^{\dagger} \end{bmatrix} R,\\
A^{(1,2,4)} =& G\begin{bmatrix} I_r & X \\ -G_2^{\dagger}G_1 & -G_2^{\dagger}G_1X \end{bmatrix} R = G\begin{bmatrix} I_r \\ -G_2^{\dagger}G_1 \end{bmatrix}\begin{bmatrix} I_r & X \end{bmatrix} R,\\
A^{(1,3,4)} =& G\begin{bmatrix} I_r & -R_1R_2^{\dagger} \\ -G_2^{\dagger}G_1 & Z \end{bmatrix} R,
\end{aligned}
$$

$$\begin{aligned} A^{\dagger} =& G\begin{bmatrix} I_r & -R_1R_2^{\dagger} \\ -G_2^{\dagger}G_1 & G_2^{\dagger}G_1R_1R_2^{\dagger} \end{bmatrix} R \\ =& G\begin{bmatrix} I_r \\ -G_2^{\dagger}G_1 \end{bmatrix}\begin{bmatrix} I_r & -R_1R_2^{\dagger} \end{bmatrix} R \\ =& (G_1 - G_2G_2^{\dagger}G_1)(R_1 - R_1R_2^{\dagger}R_2). \end{aligned}$$

Theorem 3.11.3. *If $A \in \mathbb{C}_r^{m\times n}$, and if $R = \begin{bmatrix} R_1 \\ R_2 \end{bmatrix}$ i $G = \begin{bmatrix} G_1 & G_2 \end{bmatrix}$, matrix which A transformed into normal form $A = R^{-1}\begin{bmatrix} B & \mathbb{O} \\ \mathbb{O} & \mathbb{O} \end{bmatrix} G^{-1}$ then:*

$$\begin{aligned} A^{(1)} &= G\begin{bmatrix} B^{-1} & X \\ Y & Z \end{bmatrix} R, \\ A^{(1,2)} &= G\begin{bmatrix} B^{-1} & X \\ Y & YBX \end{bmatrix} R = G\begin{bmatrix} B^{-1} \\ Y \end{bmatrix}\begin{bmatrix} I_r & BX \end{bmatrix} R, \\ A^{(1,3)} &= G\begin{bmatrix} B^{-1} & -B^{-1}R_1R_2^{\dagger} \\ Y & Z \end{bmatrix} R, \\ A^{(1,4)} &= G\begin{bmatrix} B^{-1} & X \\ -G_2^{\dagger}G_1B^{-1} & Z \end{bmatrix} R, \\ A^{(1,2,3)} &= G\begin{bmatrix} B^{-1} & -B^{-1}R_1R_2^{\dagger} \\ Y & -YR_1R_2^{\dagger} \end{bmatrix} R = G\begin{bmatrix} B^{-1} \\ Y \end{bmatrix}\begin{bmatrix} I_r, & -R_1R_2^{\dagger} \end{bmatrix} R, \\ A^{(1,2,4)} &= G\begin{bmatrix} B^{-1} & X \\ -G_2^{\dagger}G_1 & -G_2^{\dagger}G_1X \end{bmatrix} R = G\begin{bmatrix} I_r \\ -G_2^{\dagger}G_1 \end{bmatrix}\begin{bmatrix} B^{-1} & X \end{bmatrix} R, \\ A^{(1,3,4)} &= G\begin{bmatrix} B^{-1} & -B^{-1}R_1R_2^{\dagger} \\ -G_2^{\dagger}G_1B^{-1} & Z \end{bmatrix} R, \end{aligned}$$

$$A^{\dagger} = G \begin{bmatrix} B^{-1} & -B^{-1}R_1R_2^{\dagger} \\ -G_2^{\dagger}G_1B^{-1} & G_2^{\dagger}G_1R_1B^{-1}R_2^{\dagger} \end{bmatrix} R$$

$$=G \begin{bmatrix} I_r \\ -G_2^{\dagger}G_1 \end{bmatrix} B^{-1} \begin{bmatrix} I_r & -R_1R_2^{\dagger} \end{bmatrix} R,$$

$$= (G_1 - G_2G_2^{\dagger}G_1)B^{-1}(R_1 - R_1R_2^{\dagger}R_2).$$

Theorem 3.11.4. *If* $A \in \mathbb{C}_r^{m\times n}$, *and if* $R = \begin{bmatrix} R_1 \\ R_2 \end{bmatrix}$ *Regular matrices and F permutation matrix, so that the vase transformation* $A = R^{-1} \begin{bmatrix} I_r & K \\ \mathbb{O} & \mathbb{O} \end{bmatrix} F^*$ *then:*

$$A^{(1)} = F \begin{bmatrix} I_r - KY & X - KZ \\ Y & Z \end{bmatrix} R,$$

$$A^{(1,2)} = F \begin{bmatrix} I_r - KY & X - KYX \\ Y & YX \end{bmatrix} R = F \begin{bmatrix} I_r - KY \\ Y \end{bmatrix} \begin{bmatrix} I_r & X \end{bmatrix} R,$$

$$A^{(1,3)} = F \begin{bmatrix} I_r - KY & -R_1R_2^{\dagger} - KZ \\ Y & Z \end{bmatrix} R,$$

$$A^{(1,4)} = F \begin{bmatrix} (I_r + KK^*)^{-1} & X - KZ \\ K^*(I_r + KK^*)^{-1} & Z \end{bmatrix} R,$$

$$A^{(1,2,3)} = F \begin{bmatrix} I_r - KY & -(I_r - KY)B^{-1}R_1R_2^{\dagger} \\ Y & -YR_1R_2^{\dagger} \end{bmatrix} R,$$

$$= F \begin{bmatrix} I_r - KY \\ Y \end{bmatrix} \begin{bmatrix} I_r, & -R_1R_2^{\dagger} \end{bmatrix} R,$$

$$A^{(1,2,4)} = F \begin{bmatrix} (I_r + KK^*)^{-1} & (I_r + KK^*)^{-1}X \\ K^*(I_r + KK^*)^{-1} & K^*(I_r + KK^*)^{-1}X \end{bmatrix} R$$

$$= F \begin{bmatrix} I_r \\ K^* \end{bmatrix} (I_r + KK^*)^{-1} \begin{bmatrix} I_r & X \end{bmatrix} R,$$

$$A^{(1,3,4)} = F\begin{bmatrix} (I_r+KK^*)^{-1} & -R_1R_2^{\dagger}-KZ \\ K^*(I_r+KK^*)^{-1} & Z \end{bmatrix} R,$$

$$A^{\dagger} = F\begin{bmatrix} (I_r+KK^*)^{-1} & -(I_r+KK^*)^{-1}R_1R_2^{\dagger} \\ K^*(I_r+KK^*)^{-1} & -K^*(I_r+KK^*)^{-1}R_1R_2^{\dagger} \end{bmatrix} R$$

$$= F\begin{bmatrix} I_r \\ K^* \end{bmatrix}(I_r+KK^*)^{-1}\begin{bmatrix} I_r & -R_1R_2^{\dagger} \end{bmatrix} R.$$

Theorem 3.11.5. *If $A \in \mathbb{C}_r^{m\times n}$, and if $G = \begin{bmatrix} G_1 & G_2 \end{bmatrix}$ Regular matrices and E permutation matrix, so that is valid $A = E^*\begin{bmatrix} I_r & \mathbb{O} \\ K & \mathbb{O} \end{bmatrix} G^{-1}$ then:*

$$A^{(1)} = G\begin{bmatrix} I_r-XK & X \\ Y-ZK & Z \end{bmatrix} E,$$

$$A^{(1,2)} = G\begin{bmatrix} I_r-XK & X \\ Y-YXK & YX \end{bmatrix} E = G\begin{bmatrix} I_r \\ Y \end{bmatrix}\begin{bmatrix} I_r-XK, & X \end{bmatrix} E,$$

$$A^{(1,3)} = G\begin{bmatrix} (I_r+K^*K)^{-1} & (I_r+K^*K)^{-1}K^* \\ Y-ZK & Z \end{bmatrix} E,$$

$$A^{(1,4)} = G\begin{bmatrix} I_r-XK & X \\ -G_2^{\dagger}G_1-ZK & Z \end{bmatrix} E,$$

$$A^{(1,2,3)} = G\begin{bmatrix} (I_r+K^*K)^{-1} & (I_r+K^*K)^{-1}K^* \\ Y(I_r+K^*K)^{-1} & -Y(I_r+K^*K)^{-1}K^* \end{bmatrix} E$$

$$= G\begin{bmatrix} I_r \\ Y \end{bmatrix}(I_r+K^*K)^{-1}\begin{bmatrix} I_r, & K^* \end{bmatrix} E$$

$$A^{(1,2,4)} = G\begin{bmatrix} I_r-XK & X \\ -G_2^{\dagger}G_1(I_r-XK) & -G_2^{\dagger}G_1X \end{bmatrix} E$$

$$= G\begin{bmatrix} I_r \\ -G_2^{\dagger}G_1 \end{bmatrix}\begin{bmatrix} I_r-XK & X \end{bmatrix} E,$$

$$A^{(1,3,4)} = G\begin{bmatrix} (I_r+K^*K)^{-1} & (I_r+K^*K)^{-1}K^* \\ -G_2^{\dagger}G_1-ZK & Z \end{bmatrix} E,$$

$$A^{\dagger} = G \begin{bmatrix} (I_r + K^*K)^{-1} & (I_r + K^*K)^{-1}K^* \\ -G_2^{\dagger}G_1(I_r + K^*K)^{-1} & G_2^{\dagger}G_1(I_r + K^*K)^{-1}K^* \end{bmatrix} R$$

$$= G \begin{bmatrix} I_r \\ -G_2^{\dagger}G_1 \end{bmatrix} (I_r + K^*K)^{-1} \begin{bmatrix} I_r & K^* \end{bmatrix} E.$$

Theorem 3.11.6. *Let $A \in \mathbb{C}_r^{m\times n}$, and let E and F are permutation matrices, so that is valid* $A = E^* \begin{bmatrix} A_{11} & A_{12} \\ A_{21} & A_{22} \end{bmatrix} F^*, A_{11} \in \mathbb{C}_r^{r\times r}$. *Using marks*

$$L = A_{21}A_{11}^{-1}, \quad R = A_{11}^{-1}A_{12}, \quad M = I_r + RR^*, \quad N = I_r + LL^*$$

we get:

$$A^{(1)} = F \begin{bmatrix} A_{11}^{-1} - XL - DY - DZL & X \\ Y & Z \end{bmatrix} E,$$

$$A^{(1,2)} = F \begin{bmatrix} A_{11}^{-1} - XL - DY + DYA_{11}XL & X - DYA_{11}X \\ Y - YA_{11}XL & YA_{11}X \end{bmatrix} E$$

$$= F \begin{bmatrix} A_{11}^{-1} - DY \\ Y \end{bmatrix} \begin{bmatrix} I_r - A_{11}XL, & A_{11}X \end{bmatrix} E,$$

$$A^{(1,3)} = F \begin{bmatrix} (NA_{11})^{-1} - DY & (NA_{11})^{-1}L^* - DZ \\ Y & Z \end{bmatrix} E,$$

$$A^{(1,4)} = F \begin{bmatrix} (A_{11}M)^{-1} - XL & X \\ R^*(A_{11}M)^{-1} - ZL & Z \end{bmatrix} E,$$

$$A^{(1,2,3)} = F \begin{bmatrix} (NA_{11})^{-1} - DY & (NA_{11})^{-1}L^* - DYL^* \\ Y & YL^* \end{bmatrix} E$$

$$= F \begin{bmatrix} (NA_{11})^{-1} - DY \\ Y \end{bmatrix} \begin{bmatrix} I_r, & L^* \end{bmatrix} E,$$

$$A^{(1,2,4)} = F\begin{bmatrix} (A_{11}M)^{-1} - XL & X \\ D^*(A_{11}M)^{-1} - D^*XL & R^*X \end{bmatrix} E$$

$$= F\begin{bmatrix} I_r \\ D^* \end{bmatrix} \begin{bmatrix} (A_{11}M)^{-1} - XL & X \end{bmatrix} E,$$

$$A^{(1,3,4)} = F\begin{bmatrix} (NA_{11})^{-1} + (A_{11}M)^{-1} - A_{11}^{-1} + DZL & (NA_{11})^{-1}L^* - DZ \\ D^*(A_{11}M)^{-1} - ZL & Z \end{bmatrix} E,$$

$$A^{\dagger} = F\begin{bmatrix} (NA_{11}M)^{-1} & (NA_{11}M)^{-1}L^* \\ D^*(NA_{11}M)^{-1} & D^*(NA_{11}M)^{-1}L^* \end{bmatrix} E$$

$$= F\begin{bmatrix} I_r \\ D^* \end{bmatrix} (NA_{11}M)^{-1} \begin{bmatrix} I_r & L^* \end{bmatrix} E.$$

The next two theorems for block representation Moore–Penrose's inverse is introduced into Zlobec [91].

Theorem 3.11.7. *Moore–Penrose inverse of the matrix$A \in \mathbb{C}^{m\times n}$ is equal to*

$$A^{\dagger} = A^*(A^*AA^*)^{(1)}A^*.$$

Theorem 3.11.8. *Let $A \in \mathbb{C}_r^{m\times n}$ be matrix presented in the form $A = \begin{bmatrix} A_{11} & A_{12} \\ A_{21} & A_{22} \end{bmatrix}$, where $A_{11} \in \mathbb{C}_r^{r\times r}$. Then*

$$A^{\dagger} = \begin{bmatrix} A_{11} & A_{12} \end{bmatrix}^* T_{11}^* \begin{bmatrix} A_{11} \\ A_{21} \end{bmatrix}^*,$$

where

$$T_{11} = \left(\begin{bmatrix} A_{11} & A_{12} \end{bmatrix} A^* \begin{bmatrix} A_{11} \\ A_{21} \end{bmatrix} \right)^{-1}.$$

It can also be used Hermite algorithm, which is based on the next formula

$$A^{\dagger} = A^*(AA^*AA^*)^{(1,2)}A^* = \left((AA^*)^2\right)^{(1,2)}.$$

At the same time, to calculate the inverse $\left((AA^*)^2\right)^{(1,2)}$ can be used, for example, the block representation from Theorems 2.10.2, for $X = Y = \mathbb{O}$. If $(AA^*)^2 = R^T I_r G$, then $\left((AA^*)^2\right)^{(1,2)} = G^T I_r R = (G^T)_r^{|r} R_{|r}$, where $(G^T)^{|r}$ and $R_{|r}$ represent the first r column matrix G^T and r the first type of matrix R, respectively. Also known as the bloc representation group and Drazin's inverse. Next theorem is given in the paper [50].

Theorem 3.11.9. *Let $A \in \mathbb{C}_r^{n\times n}$, and R be regular matrix, such that the following is valid* $RAR^{-1} = \begin{bmatrix} T_1 & T_2 \\ \mathbb{O} & \mathbb{O}. \end{bmatrix}$. *Then*

$$A^{\#} = R^{-1} \begin{bmatrix} T_1^{-1} & T_1^{-2}T_2 \\ \mathbb{O} & \mathbb{O} \end{bmatrix} R.$$

This result is generalized Hartwig, which has developed a method for calculating Drazin's inverse [23], described by the following theorem.

Theorem 3.11.10. *If $A \in \mathbb{C}_r^{n\times n}$, and R is nonsingular matrix, such that* $RAR^{-1} = \begin{bmatrix} U & V \\ \mathbb{O} & \mathbb{O} \end{bmatrix}$. *Then*

$$A^D = R^{-1} \begin{bmatrix} U_d & U_d^2 V_d \\ \mathbb{O} & \mathbb{O} \end{bmatrix} R.$$

If X $XAX = X$, it follows that there is no integer $t \in \{1,\dots,r\}$, where $r = rank(A)$, such that X can now present in the form of

$$X = W_1(W_2AW_1)^{-1}W_2, \quad W_1 \in \mathbb{C}^{n\times t}, \quad W_2 \in \mathbb{C}^{t\times m}, \quad rank(W_2AW_1) = t. \tag{3.11.1}$$

Based on this, we have the following assertion of the block representation $\{1,2,3\}$ i $\{1,2,4\}$ inverse from Ref. [62].

Theorem 3.11.11. *Let $A \in \mathbb{C}_r^{m\times n}$ and let matrices W_1 and W_2 satisfies conditions from* (3.11.1). *Then it can get the next block representation* $\{1,2,3\}$

i $\{1,2,4\}$ *generalized inverses for A, under the condition that* $(G^i_{1,2,3})$ *i* $(G^i_{1,2,4})$ *correspond to* (T_i), $i \in \{1,\ldots 11\}$*:*

$$(G^1_{1,2,3}) \quad A\{1,2,3\} = G \begin{bmatrix} U_1 \\ U_2 \end{bmatrix} U_1^{-1} \left((RR^*)^{-1}\big|^{r}_{r} \right)^{-1} \begin{bmatrix} I_r & \mathbb{O} \end{bmatrix} (R^*)^{-1}$$

$$= G \begin{bmatrix} U_1 \\ U_2 \end{bmatrix} \left((RR^*)^{-1}\big|^{r}_{r} U_1 \right)^{-1} (R^*)^{-1}{}_{|r},$$

$$(G^1_{1,2,4}) \quad A\{1,2,4\} = (G^*)^{-1} \begin{bmatrix} I_r \\ \mathbb{O} \end{bmatrix} \left((G^*G)^{-1}\big|^{r}_{r} \right)^{-1} V_1^{-1} \begin{bmatrix} V_1 \\ V_2 \end{bmatrix} R$$

$$= (G^*)^{-1}\big|^{r} \left(V_1 (G^*G)^{-1}\big|^{r}_{r} \right)^{-1} \begin{bmatrix} V_1 \\ V_2 \end{bmatrix} R,$$

$$(G^2_{1,2,3}) \quad A\{1,2,3\} = G \begin{bmatrix} U_1 \\ U_2 \end{bmatrix} U_1^{-1} \left(B^* (RR^*)^{-1}\big|^{r}_{r} B \right)^{-1} \begin{bmatrix} B^*, & \mathbb{O} \end{bmatrix} (R^*)^{-1}$$

$$= G \begin{bmatrix} U_1 \\ U_2 \end{bmatrix} \left((RR^*)^{-1}\big|^{r}_{r} B U_1 \right)^{-1} (R^*)^{-1}{}_{|r},$$

$$(G^2_{1,2,4}) \quad A\{1,2,4\} = (G^*)^{-1} \begin{bmatrix} I_r \\ \mathbb{O} \end{bmatrix} \left((G^*G)^{-1}\big|^{r}_{r} \right)^{-1} (V_1 B)^{-1} \begin{bmatrix} V_1, & B \end{bmatrix} R$$

$$= (G^*)^{-1}\big|^{r} \left(V_1 B (G^*G)^{-1}\big|^{r}_{r} \right)^{-1} \begin{bmatrix} V_1 & B \end{bmatrix} R,$$

$$(G^3_{1,2,3}) \quad A\{1,2,3\} = F \begin{bmatrix} S_1 \\ S_2 \end{bmatrix} (S_1 + KS_2)^{-1} \left((RR^*)^{-1}\big|^{r}_{r} \right)^{-1} \begin{bmatrix} I_r & \mathbb{O} \end{bmatrix} (R^*)^{-1}$$

$$= F \begin{bmatrix} S_1 \\ S_2 \end{bmatrix} \left((RR^*)^{-1}\big|^{r}_{r} (S_1 + KS_2) \right)^{-1} (R^*)^{-1}{}_{|r},$$

$$(G^3_{1,2,4}) \quad A\{1,2,4\} = F \begin{bmatrix} I_r \\ K^* \end{bmatrix} (I_r + KK^*)^{-1} V_1^{-1} \begin{bmatrix} V_1, & V_2 \end{bmatrix} R$$

$$= \left(F^{|r} + F^{n-r|} K^* \right) (V_1 (I_r + KK^*))^{-1} \begin{bmatrix} V_1, & V_2 \end{bmatrix} R,$$

$$(G^4_{1,2,3}) \quad A\{1,2,3\} = G \begin{bmatrix} U_1 \\ U_2 \end{bmatrix} U_1^{-1} (I_r + KK^*)^{-1} \begin{bmatrix} I_r, & K^* \end{bmatrix} E$$

$$= G \begin{bmatrix} U_1 \\ U_2 \end{bmatrix} ((I_r + KK^*) U_1)^{-1} \left(E_{|r} + K^* E_{n-r|} \right),$$

$$(G^4_{1,2,4})\,A\{1,2,4\} = (G^*)^{-1}\begin{bmatrix} I_r \\ \mathbb{O} \end{bmatrix}\left((G^*G)^{-1}\big|^{r}_{|r}\right)^{-1}(T_1+T_2K)^{-1}\begin{bmatrix} T_1, & T_2 \end{bmatrix}E$$

$$= (G^*)^{-1|r}\left((T_1+T_2K)(G^*G)^{-1}\big|^{r}_{|r}\right)^{-1}\begin{bmatrix} T_1, & T_2 \end{bmatrix}E,$$

$$(G^5_{1,2,3}) \quad A\{1,2,3\} = G\begin{bmatrix} U_1 \\ U_2 \end{bmatrix}U_1^{-1}\begin{bmatrix} I_r & \mathbb{O} \end{bmatrix}U = G\begin{bmatrix} U_1 \\ U_2 \end{bmatrix}U_1^{-1}U_{|r},$$

$$(G^5_{1,2,4}) \quad A\{1,2,4\} = (G^*)^{-1}\begin{bmatrix} I_r \\ \mathbb{O} \end{bmatrix}\left((G^*G)^{-1}\big|^{r}_{|r}\right)^{-1}D_1^{-1}\begin{bmatrix} D_1 & D_2 \end{bmatrix}U$$

$$= (G^*)^{-1|r}\left(D_1(G^*G)^{-1}\big|^{r}_{|r}\right)^{-1}\begin{bmatrix} D_1 & D_2 \end{bmatrix}U,$$

$$(G^6_{1,2,3}) \quad A\{1,2,3\} = V\begin{bmatrix} C_1 \\ C_2 \end{bmatrix}C_1^{-1}\left((RR^*)^{-1}\big|^{r}_{|r}\right)^{-1}\begin{bmatrix} I_r, & \mathbb{O} \end{bmatrix}(R^*)^{-1}$$

$$= V\begin{bmatrix} C_1 \\ C_2 \end{bmatrix}\left((RR^*)^{-1}\big|^{r}_{|r}C_1\right)^{-1}(R^*)^{-1}{}_{|r},$$

$$(G^6_{1,2,4}) \quad A\{1,2,4\} = V\begin{bmatrix} I_r \\ \mathbb{O} \end{bmatrix}V_1^{-1}\begin{bmatrix} V_1, & V_2 \end{bmatrix}R = V^{|r}V_1^{-1}\begin{bmatrix} V_1, & V_2 \end{bmatrix}R,$$

$$(G^7_{1,2,3}) \quad A\{1,2,3\} = V\begin{bmatrix} C_1 \\ C_2 \end{bmatrix}C_1^{-1}(B^*B)^{-1}\begin{bmatrix} B^*, & \mathbb{O} \end{bmatrix}U$$

$$= V\begin{bmatrix} C_1 \\ C_2 \end{bmatrix}(BC_1)^{-1}U_{|r},$$

$$(G^7_{1,2,4}) \quad A\{1,2,4\} = V\begin{bmatrix} I_r \\ \mathbb{O} \end{bmatrix}(D_1B)^{-1}\begin{bmatrix} D_1 & D_2 \end{bmatrix}U = V^{|r}(D_1B)^{-1}\begin{bmatrix} D_1 & D_2 \end{bmatrix}U,$$

$$(G^8_{1,2,3}) \quad A\{1,2,3\} = F\begin{bmatrix} S_1 \\ S_2 \end{bmatrix}(BS_1+KS_2)^{-1}\begin{bmatrix} I_r, & \mathbb{O} \end{bmatrix}U$$

$$= F\begin{bmatrix} S_1 \\ S_2 \end{bmatrix}(BS_1+KS_2)^{-1}U_{|r},$$

$$(G^8_{1,2,4}) \quad A\{1,2,4\} = F\begin{bmatrix} B^* \\ K^* \end{bmatrix}(D_1(BB^*+KK^*))^{-1}\begin{bmatrix} D_1, & D_2 \end{bmatrix}U,$$

$$(G^9_{1,2,3}) \quad A\{1,2,3\} = V\begin{bmatrix} C_1 \\ C_2 \end{bmatrix}((B^*B+K^*K)C_1)^{-1}\begin{bmatrix} B^*, & K^* \end{bmatrix}E,$$

$$(G^9_{1,2,4})\quad A\{1,2,4\} = V\begin{bmatrix} I_r \\ \mathbb{O} \end{bmatrix}(T_1B+T_2K)^{-1}\begin{bmatrix} T_1, & T_2 \end{bmatrix}E$$

$$= V^{|r}(T_1B+T_2K)^{-1}\begin{bmatrix} T_1, & T_2 \end{bmatrix}E,$$

$$(G^{10a}_{1,2,3})\, A\{1,2,3\} = F\begin{bmatrix} S_1 \\ S_2 \end{bmatrix}(S_1+TS_2)^{-1}A_{11}^{-1}(I_r+S^*S)^{-1}\begin{bmatrix} I_r, & S^* \end{bmatrix}E$$

$$= F\begin{bmatrix} S_1 \\ S_2 \end{bmatrix}((I_r+S^*S)A_{11}(S_1+TS_2))^{-1}(E_{|r}+S^*E_{n-r|}),$$

$$(G^{10a}_{1,2,4})\, A\{1,2,4\} = F\begin{bmatrix} I_r \\ T^* \end{bmatrix}(I_r+TT^*)^{-1}A_{11}^{-1}(T_1+T_2S)^{-1}\begin{bmatrix} T_1, & T_2 \end{bmatrix}E$$

$$= \left(F^{|r}+F^{n-r|}T^*\right)((T_1+T_2S)A_{11}(I_r+TT^*))^{-1}\begin{bmatrix} T_1, & T_2 \end{bmatrix}E,$$

$$(G^{10b}_{1,2,3})\, A\{1,2,3\} = F\begin{bmatrix} S_1 \\ S_2 \end{bmatrix}(A_{11}S_1+A_{12}S_2)^{-1}A_{11}(A_{11}^*A_{11}+A_{21}^*A_{21})^{-1}\begin{bmatrix} A_{11}^*, & A_{21}^* \end{bmatrix}E,$$

$$(G^{10b}_{1,2,4})\, A\{1,2,4\} = F\begin{bmatrix} A_{11}^* \\ A_{12}^* \end{bmatrix}(A_{11}A_{11}^*+A_{12}A_{12}^*)^{-1}A_{11}(T_1A_{11}+T_2A_{21})^{-1}\begin{bmatrix} T_1, & T_2 \end{bmatrix}E,$$

$$(G^{11}_{1,2,3})A\{1,2,3\} = R^{\;1}\begin{bmatrix} Q_1 \\ Q_2 \end{bmatrix}(Q_1+T_1^{-1}T_2Q_2)^{-1}\left(T_1^*(RR^*)^{-1}{}^{|r}_{|r}T_1\right)^{-1}\begin{bmatrix} T_1^* \\ \mathbb{O} \end{bmatrix}(R^{-1})^*$$

$$= R^{-1}\begin{bmatrix} Q_1 \\ Q_2 \end{bmatrix}\left(T_1^*(RR^*)^{-1}{}^{|r}_{|r}(T_1Q_1+T_2Q_2)\right)^{-1}(R^{-1|r}T_1)^*,$$

$$(G^{11}_{1,2,4})A\{1,2,4\} = R^*\begin{bmatrix} I_r \\ (T_1^{-1}T_2)^* \end{bmatrix}\left(V_1\begin{bmatrix} T_1, & T_2 \end{bmatrix}RR^*\begin{bmatrix} I_r \\ (T_1^{-1}T_2)^* \end{bmatrix}\right)^{-1}\begin{bmatrix} V_1, & V_2 \end{bmatrix}R.$$

Follows the implementation of the said bloc representation of generalized inverse. Here are just a tool to determine representation $(G^i_{1,2,3})$ i $(G^i_{1,2,4})$, $i \in \{1,3,4\}$.

```
G1231[A_List,R1_,R2_]:=
Block[A=b,W1=R1,W2=R2, ran,r,g,u1,u2,rz,pom1,pom2,
   {r,g}=t1[A]; ran=rank[A];
   pom1=Inverse[g].W1; pom2=W2.Inverse[r];
   u1=Take[pom1,ran];
   u2=Drop[pom1,ran];
   rz=Inverse[r.Hermit[r]];
   rz=Take[rz,ran];
```

```
   rz=Transpose[Take[Transpose[rz],ran]];
   Return[g.pom1.Inverse[rz.u1].Take[Inverse[Hermit[r]],ran]];
]

G1241[A_List,R1_,R2_]:=
Block[W1=R1,W2=R2, ran,r,g,v1,v2,gz,gz1,pom2,
   {r,g}=t1[A]; ran=rank[A];
   pom2=W2.Inverse[r];
   v1=Transpose[Take[Transpose[pom2],ran]];
   v2=Transpose[Drop[Transpose[pom2],ran]];
   gz=Inverse[Hermit[g].g];
   gz=Take[gz,ran];
   gz=Transpose[Take[Transpose[gz],ran]];
   gz1=Transpose[Take[Transpose[Inverse[Hermit[g]]],ran]];
   Return[gz1.Inverse[v1.gz].pom2.r];
]

G1233[A_List,R1_,R2_]:=
Block[{W1=R1,W2=R2, ran,r,f,s1,s2,k,v1,v2,pom1,pom3},
   {r,f}=t3[A]; ran=rank[A];
   pom1=Hermit[f].W1;
   s1=Take[pom1,ran];
   s2=Drop[pom1,ran];
   pom3=Take[r.A.f,ran];
   k=Transpose[Drop[Transpose[pom3],ran]];
   rz=Inverse[r.Hermit[r]];
   rz=Take[rz,ran];
   rz=Transpose[Take[Transpose[rz],ran]];
   Return[f.pom1.Inverse[rz.(s1+k.s2)].Take[Inverse[Hermit[r]],ran]];
]

G1243[A_List,R1_,R2_]:=
Block[{W1=R1,W2=R2, ran,r,f,v1,v2,k,pom2,pom3},
   {r,f}=t3[A]; ran=rank[A];
   pom2=W2.Inverse[r];
   v1=Transpose[Take[Transpose[pom2],ran]];
   v2=Transpose[Drop[Transpose[pom2],ran]];
   pom3=Take[r.A.f,ran];
```

```
    k=Transpose[Drop[Transpose[pom3],ran]];
    Return[f.Hermit[pom3].Inverse[IdentityMatrix[ran]+k.Hermit[k]].
    Inverse[v1].pom2.r];
]

G1234[A_List,R1_,R2_]:=
Block[{W1=R1,W2=R2, ran,e,g,u1,u2,k,pom1,pom3},
    {e,g}=t4[A]; ran=rank[A];
    pom1=Inverse[g].W1;
    u1=Take[pom1,ran];
    u2=Drop[pom1,ran];
    pom3=Drop[e.A.g,ran];
    k=Transpose[Take[Transpose[pom3],ran]];
    pom3=Conjugate[Take[Transpose[e.A.g],ran]];
    Return[g.pom1.Inverse[(IdentityMatrix[ran]+k.Hermit[k]).u1].pom3.e];
]

G1244[A_List,R1_,R2_]:=
Block[{W1=R1,W2=R2, ran,e,g,u1,u2,k,gz,t1,t2,pom2,pom3},
    {e,g}=t4[A]; ran=rank[A];
    pom2=W2.Hermit[e];
    t1=Transpose[Take[Transpose[pom2],ran]];
    t2=Transpose[Drop[Transpose[pom2],ran]];
    pom3=Drop[e.A.g,ran];
    k=Transpose[Take[Transpose[pom3],ran]];
    gz=Inverse[Hermit[g].g];
    gz=Take[gz,ran];
    gz=Transpose[Take[Transpose[gz],ran]];
    gz1=Transpose[Take[Transpose[Inverse[Hermit[g]]],ran]];
    Return[gz1.Inverse[(t1+t2.k).gz].pom2.e];

]
```

Example 3.11.3. Using

$$A = \begin{bmatrix} -1 & 0 & 1 & 2 \\ -1 & 1 & 0 & -1 \\ 0 & -1 & 1 & 3 \\ 0 & 1 & -1 & -3 \\ 1 & -1 & 0 & 1 \\ 1 & 0 & -1 & -2 \end{bmatrix}, \quad W_1 = \begin{bmatrix} 2 & 0 \\ 0 & 1 \\ 1 & 0 \\ 4 & 2 \end{bmatrix}$$

$$W_2 = \begin{bmatrix} 3 & 1 & 3 & 1 & 2 & -1 \\ 0 & -1 & 0 & 0 & -2 & -1 \end{bmatrix}$$

expression `x=G1231[a,w1,w2]` we get value

$$X = \begin{bmatrix} -\frac{2}{17} & -\frac{3}{17} & \frac{1}{17} & -\frac{1}{17} & \frac{3}{17} & \frac{2}{17} \\ \frac{19}{102} & \frac{10}{51} & -\frac{1}{102} & \frac{1}{102} & -\frac{10}{51} & -\frac{19}{102} \\ -\frac{1}{17} & -\frac{3}{34} & \frac{1}{34} & -\frac{1}{34} & \frac{3}{34} & \frac{1}{17} \\ \frac{7}{51} & \frac{2}{51} & \frac{5}{51} & -\frac{5}{51} & -\frac{2}{51} & -\frac{7}{51} \end{bmatrix}$$

In the case $W_1 = Q^*$, $W_2 = P^*$ we can obtain the following representation of Moore–Penrose's inverse.

Corollary 3.11.12. *Moore–Penrose's inverse of given matrix* $A \in \mathbb{C}_r^{m\times n}$ *can be represented as follows, where representation* $(A_\dagger^i)$ *corresponds to bloc decompositions*

(T_i), $i \in \{1,\ldots,9,10a,10b,11\}$

$$(A_\dagger^1) \quad A^\dagger = \left(G^{-1}{}_{|r}\right)^* \left(\left(R^{-1^{|r}}\right)^* A \left(G^{-1}{}_{|r}\right)^*\right)^{-1} \left(R^{-1^{|r}}\right)^*$$

$$= \left(G^{-1}{}_{|r}\right)^* \left((RR^*)^{-1}{}_{|r}^{|r}(G^*G)^{-1}{}_{|r}^{|r}\right)^{-1} \left(R^{-1^{|r}}\right)^*,$$

$$(A_\dagger^2) \quad A^\dagger = \left(G^{-1}{}_{|r}\right)^* \left(\left(R^{-1^{|r}}B\right)^* A \left(G^{-1}{}_{|r}\right)^*\right)^{-1} \left(R^{-1^{|r}}B\right)^*$$

$$= \left(G^{-1}{}_{|r}\right)^* \left(B^*(RR^*)^{-1}{}_{|r}^{|r}B(G^*G)^{-1}{}_{|r}^{|r}\right)^{-1} B^* \left(R^{-1^{|r}}\right)^*,$$

$$(A_{\dagger}^{3})\quad A^{\dagger}=F\begin{bmatrix} I_r \\ K^* \end{bmatrix}\left(\left(R^{-1|_r}\right)^* AF\begin{bmatrix} I_r \\ K^* \end{bmatrix}\right)^{-1}\left(R^{-1|_r}\right)^*$$

$$=\left(F^{|r}+F^{n-r|}K\right)\left((RR^*)^{-1}{}^{|r}_{|r}(I_r+KK^*)\right)^{-1}\left(R^{-1|_r}\right)^*,$$

$$(A_{\dagger}^{4})\quad A^{\dagger}=(G^{-1}{}_{|r})^*\left(\begin{bmatrix} I_r, & K^* \end{bmatrix}EA\,(G^{-1}{}_{|r})^*\right)^{-1}\begin{bmatrix} I_r, & K^* \end{bmatrix}E$$

$$=(G^{-1}{}_{|r})^*\left((I_r+K^*K)(G^*G)^{-1}{}^{|r}_{|r}\right)^{-1}\left(E_{|r}+K^*E_{n-r|}\right),$$

$$(A_{\dagger}^{5})\quad A^{\dagger}=(G^{-1}{}_{|r})^*\left(U_{|r}A\,(G^{-1}{}_{|r})^*\right)^{-1}U_{|r}=(G^{-1}{}_{|r})^*\left((G^*G)^{-1}{}^{|r}_{|r}\right)^{-1}U_{|r},$$

$$(A_{\dagger}^{6})\quad A^{\dagger}=V^{|r}\left(\left(R^{-1|_r}\right)^* AV^{|r}\right)^{-1}\left(R^{-1|_r}\right)^*=V^{|r}\left((RR^*)^{-1}{}^{|r}_{|r}\right)^{-1}\left(R^{-1|_r}\right)^*,$$

$$(A_{\dagger}^{7})\quad A^{\dagger}=V^{|r}\left(B^*U_{|r}AV^{|r}\right)^{-1}B^*U_{|r}=V^{|r}B^{-1}U_{|r},$$

$$(A_{\dagger}^{8})\quad A^{\dagger}=F\begin{bmatrix} B^* \\ K^* \end{bmatrix}\left(U_{|r}AF\begin{bmatrix} B^* \\ K^* \end{bmatrix}\right)^{-1}U_{|r}=F\begin{bmatrix} B^* \\ K^* \end{bmatrix}(BB^*+KK^*)^{-1}U_{|r},$$

$$(A_{\dagger}^{9})\quad A^{\dagger}=V^{|r}\left(\begin{bmatrix} B^* & K^* \end{bmatrix}EAV^{|r}\right)^{-1}\begin{bmatrix} B^* & K^* \end{bmatrix}E$$

$$=V^{|r}\,(B^*B+K^*K)^{-1}\begin{bmatrix} B^* & K^* \end{bmatrix}E,$$

$$(A_{\dagger}^{10a})\,A^{\dagger}=F\begin{bmatrix} I_r \\ T^* \end{bmatrix}\left(A_{11}^*\begin{bmatrix} I_r, & S^* \end{bmatrix}EAF\begin{bmatrix} I_r \\ T^* \end{bmatrix}\right)^{-1}A_{11}^*\begin{bmatrix} I_r, & S^* \end{bmatrix}E$$

$$=\left(F^{|r}+F^{n-r|}T^*\right)\left((I_r+S^*S)A_{11}(I_r+TT^*)\right)^{-1}\left(E_{|r}+S^*E_{n-r|}\right),$$

$$(A_{\dagger}^{10b})\,A^{\dagger}=F\begin{bmatrix} A_{11}^* \\ A_{12}^* \end{bmatrix}\left((A_{11}^*)^{-1}\begin{bmatrix} A_{11}^*, & A_{21}^* \end{bmatrix}EAF\begin{bmatrix} A_{11}^* \\ A_{12}^* \end{bmatrix}\right)^{-1}(A_{11}^*)^{-1}\begin{bmatrix} A_{11}^*, & A_{21}^* \end{bmatrix}E$$

$$=F\begin{bmatrix} A_{11}^* \\ A_{12}^* \end{bmatrix}(A_{11}A_{11}^*+A_{12}A_{12}^*)^{-1}A_{11}(A_{11}^*A_{11}+A_{21}^*A_{21})^{-1}\begin{bmatrix} A_{11}^*, & A_{21}^* \end{bmatrix}E,$$

$$(A_{\dagger}^{11})\quad A^{\dagger}=R^*\begin{bmatrix} I_r \\ (T_1^{-1}T_2)^* \end{bmatrix}\left(\left(R^{-1|_r}T_1\right)^* AR^*\begin{bmatrix} I_r \\ (T_1^{-1}T_2)^* \end{bmatrix}\right)^{-1}\left(R^{-1|_r}T_1\right)^*$$

$$=R^*\begin{bmatrix} I_r \\ (T_1^{-1}T_2)^* \end{bmatrix}\left(T_1^*(RR^*)^{-1}{}^{|r}_{|r}\begin{bmatrix} T_1, & T_2 \end{bmatrix}RR^*\begin{bmatrix} I_r \\ (T_1^{-1}T_2)^* \end{bmatrix}\right)^{-1}\left(R^{-1|_r}T_1\right)^*.$$

Implementing procedures for calculating the Moore–Penrose's inverse, based on representations $(A_\dagger^1)$, $(A_\dagger^3)$ i $(A_\dagger^4)$ are given with the following codes:

```
MPt1[A_List]:=
   Block[{r,g,ran,pt,qt},
   {r,g}=t1[A];    ran=rank[A];
   qt=Hermit[Take[Inverse[g],ran]];
   pt=Conjugate[Take[Transpose[Inverse[r]],ran]];
   qt.Inverse[pt.A.qt].pt
   ];MatrixQ[A]

MPt3[A_List]:=
   Block[{r,f,ran,pt,qt},
   {r,f}=t3[A];    ran=rank[A];
   qt=f.Hermit[Take[r.A.f,ran]];
   pt=Conjugate[Take[Transpose[Inverse[r]],ran]];
   qt.Inverse[pt.A.qt].pt
   ];MatrixQ[A]

MPt4[A_List]:=
   Block[{e,g,ran,pt,qt},
   {e,g}=t4[A];    ran=rank[A];
   qt=Hermit[Take[Inverse[g],ran]];
   pt=Conjugate[Take[Transpose[e.A.g],ran]].e;
   qt.Inverse[pt.A.qt].pt
   ];MatrixQ[A]
```

Theorem 3.11.13. *Weight Moore–Penrose's inverse $A^\dagger_{M\circ,\circ N}$ for $A \in \mathbb{C}_r^{m\times n}$ has representation next block, where the representation (Z_i) follows from block decomposition (T_i), $i \in \{1,\dots,9,10a,10b,11\}$:*

$$(Z_1)\quad N\left(G^{-1}{}_{|r}\right)^*\left(\left(R^{-1^{|r}}\right)^* MAN\left(G^{-1}{}_{|r}\right)^*\right)^{-1}\left(R^{-1^{|r}}\right)^* M,$$

$$(Z_2)\quad N\left(G^{-1}{}_{|r}\right)^*\left(\left(R^{-1^{|r}}B\right)^* MAN\left(G^{-1}{}_{|r}\right)^*\right)^{-1}\left(R^{-1^{|r}}B\right)^* M,$$

$$(Z_3)\quad NF\begin{bmatrix} I_r \\ K^* \end{bmatrix}\left(\left(R^{-1^{|r}}\right)^* MANF\begin{bmatrix} I_r \\ K^* \end{bmatrix}\right)^{-1}\left(R^{-1^{|r}}\right)^* M,$$

$(Z_4)\quad N\left(G^{-1}{}_{|r}\right)^*\left(\begin{bmatrix} I_r, & K^* \end{bmatrix} EMAN\left(G^{-1}{}_{|r}\right)^*\right)^{-1}\begin{bmatrix} I_r, & K^* \end{bmatrix} EM,$

$(Z_5)\quad N\left(G^{-1}{}_{|r}\right)^*\left(U_{|r}MAN\left(G^{-1}{}_{|r}\right)^*\right)^{-1}U_{|r}M,$

$(Z_6)\quad NV^{|r}\left(\left(R^{-1^{|r}}\right)^* MANV^{|r}\right)^{-1}\left(R^{-1^{|r}}\right)^* M,$

$(Z_7)\quad NV^{|r}\left(B^*U_{|r}MANV^{|r}\right)^{-1}B^*U_{|r}M,$

$(Z_8)\quad NF\begin{bmatrix} B^* \\ K^* \end{bmatrix}\left(U_{|r}MANF\begin{bmatrix} B^* \\ K^* \end{bmatrix}\right)^{-1}U_{|r}M,$

$(Z_9)\quad NV^{|r}\left(\begin{bmatrix} B^* & K^* \end{bmatrix} EMANV^{|r}\right)^{-1}\begin{bmatrix} B^* & K^* \end{bmatrix} EM,$

$(Z_{10a})\; NF\begin{bmatrix} I_r \\ T^* \end{bmatrix}\left(A_{11}^*\begin{bmatrix} I_r, & S^* \end{bmatrix} EMANF\begin{bmatrix} I_r \\ T^* \end{bmatrix}\right)^{-1} A_{11}^*\begin{bmatrix} I_r, & S^* \end{bmatrix} EM,$

$(Z_{10b})\; NF\begin{bmatrix} A_{11}^* \\ A_{12}^* \end{bmatrix}\left((A_{11}^*)^{-1}\begin{bmatrix} A_{11}^*, & A_{21}^* \end{bmatrix} EMANF\begin{bmatrix} A_{11}^* \\ A_{12}^* \end{bmatrix}\right)^{-1}(A_{11}^*)^{-1}\begin{bmatrix} A_{11}^*, & A_{21}^* \end{bmatrix} EM,$

$(Z_{11})\quad NR^*\begin{bmatrix} I_r \\ (T_1^{-1}T_2)^* \end{bmatrix}\left(\left(R^{-1^{|r}}T_1\right)^* MANR^*\begin{bmatrix} I_r \\ (T_1^{-1}T_2)^* \end{bmatrix}!\right)^{-1}\left(R^{-1^{|r}}T_1\right)^* M.$

Proof: From $A^{\dagger}_{M\circ,\circ N} = (QN)^*(Q(QN)^*)^{-1}((MP)^*P)^{-1}(MP)^*$ follows as $A^{\dagger}_{M\circ,\circ N}$ represents an element of class $A\{1,2\}$ which satisfies

$$W_1 = NQ^*, \qquad W_2 = P^*M. \qquad \square$$

The following representation can be obtained from the main feature of weight Moore–Penrose's and inverse theorems 1.3.3.

Corollary 3.11.14. *Weight Moore–Penrose's inverse* $A^{\dagger}_{\varphi(M,N)}$ *and* $A \in \mathbb{C}_r^{m\times n}$ *can be presented as follows:*

$(W_1)\; N^{[-1]}\left(G^{-1}{}_{|r}\right)^*\left(\left(R^{-1^{|r}}\right)^* M^{[-1]}AN^{[-1]}\left(G^{-1}{}_{|r}\right)^*\right)^{-1}\left(R^{-1^{|r}}\right)^* M^{[-1]},$

$(W_2)\; N^{[-1]}\left(G^{-1}{}_{|r}\right)^*\left(\left(R^{-1^{|r}}B\right)^* M^{[-1]}AN^{[-1]}\left(G^{-1}{}_{|r}\right)^*\right)^{-1}\left(R^{-1^{|r}}B\right)^* M^{[-1]},$

$(W_3)\; N^{[-1]}F\begin{bmatrix} I_r \\ K^* \end{bmatrix}\left(\left(R^{-1^{|r}}\right)^* M^{[-1]}AN^{[-1]}F\begin{bmatrix} I_r \\ K^* \end{bmatrix}\right)^{-1}\left(R^{-1^{|r}}\right)^* M^{[-1]},$

$$(W_4)\ N^{[-1]}(G^{-1}{}_{|r})^*\left(\begin{bmatrix} I_r & K^* \end{bmatrix} EM^{[-1]}AN^{[-1]}(G^{-1}{}_{|r})^*\right)^{-1}\begin{bmatrix} I_r & K^* \end{bmatrix} EM^{[-1]},$$

$$(W_5)\ N^{[-1]}\left(G^{-1}{}_{|r}\right)^*\left(U_{|r}M^{[-1]}AN^{[-1]}\left(G^{-1}{}_{|r}\right)^*\right)^{-1}U_{|r}M^{[-1]},$$

$$(W_6)\ N^{[-1]}V^{|r}\left(\left(R^{-1^{|r}}\right)^* M^{[-1]}AN^{[-1]}V^{|r}\right)^{-1}\left(R^{-1^{|r}}\right)^* M^{[-1]},$$

$$(W_7)\ N^{[-1]}V^{|r}\left(B^*U_{|r}M^{[-1]}AN^{[-1]}V^{|r}\right)^{-1}B^*U_{|r}M^{[-1]},$$

$$(W_8)\ N^{[-1]}F\begin{bmatrix} B^* \\ K^* \end{bmatrix}\left(U_{|r}M^{[-1]}AN^{[-1]}F\begin{bmatrix} B^* \\ K^* \end{bmatrix}\right)^{-1}U_{|r}M^{[-1]},$$

$$(W_9)\ N^{[-1]}V^{|r}\left(\begin{bmatrix} B^*, & K^* \end{bmatrix} EM^{[-1]}AN^{[-1]}V^{|r}\right)^{-1}\begin{bmatrix} B^*, & K^* \end{bmatrix} EM^{[-1]},$$

$$(W_{10a})\ N^{[-1]}F\begin{bmatrix} I_r \\ T^* \end{bmatrix}\left(A_{11}^*\begin{bmatrix} I_r, & S^* \end{bmatrix} EM^{[-1]}AN^{[-1]}F\begin{bmatrix} I_r \\ T^* \end{bmatrix}\right)^{-1} A_{11}^*\begin{bmatrix} I_r, & S^* \end{bmatrix} EM^{[-1]},$$

$$(W_{10b})\ N^{[-1]}F\begin{bmatrix} A_{11}^* \\ A_{12}^* \end{bmatrix}\left((A_{11}^*)^{-1}\begin{bmatrix} A_{11}^*, & A_{21}^* \end{bmatrix} EM^{[-1]}AN^{[-1]}F\begin{bmatrix} A_{11}^* \\ A_{12}^* \end{bmatrix}\right)^{-1}$$
$$\times (A_{11}^*)^{-1}\begin{bmatrix} A_{11}^*, & A_{21}^* \end{bmatrix} EM^{[-1]},$$

$$(W_{11})\ N^{[-1]}R^*\begin{bmatrix} I_r \\ (T_1^{-1}T_2)^* \end{bmatrix}\left(\left(R^{-1^{|r}}T_1\right)^* M^{[-1]}AN^{[-1]}R^*\begin{bmatrix} I_r \\ (T_1^{-1}T_2)^* \end{bmatrix}\right)^{-1}\left(R^{-1^{|r}}T_1\right)^* M^{[-1]}.$$

Theorem 3.11.15. *Let A square matrix of order n and rank r. Then there is the group inverse if and only if the following conditions (E_i), dependent on the corresponding blocks of the decomposition (T_I) fulfilled:*

(E_1) $\quad G^{-1}{}_{|r}R^{-1^{|r}} = (RG)^{-1}{}^{|r}_{|r}$ *is invertible,*

(E_2) $\quad G^{-1}{}_{|r}R^{-1^{|r}}B = (RG)^{-1}{}^{|r}_{|r}B$ *is invertible,*

(E_3) $\quad \begin{bmatrix} I_r & K \end{bmatrix}(RF)^{-1|r} = (RF)^{-1}{}^{|r}_{|r} + K(RF)^{-1}{}^{|r}_{n-r|}$ *is invertible,*

(E_4) $\quad (EG)^{-1}{}_{|r}\begin{bmatrix} I_r \\ K \end{bmatrix} = (EG)^{-1}{}^{|r}_{|r} + (EG)^{-1}{}^{n-r|}_{|r}K$ *is invertible,*

(E_5) $U^{*|r}G^{-1}{}_{|r} = (UG)^{-1}{}^{|r}_{|r}$ *is invertible,*

(E_6) $V^{*}{}_{|r}R^{-1|r} = (RV)^{-1}{}^{|r}_{|r}$ *is invertible,*

(E_7) $V^{*}{}_{|r}U^{*|r}B = (UV)^{*}{}^{|r}_{|r}B$ *is invertible,*

(E_8) $\begin{bmatrix} B, & K \end{bmatrix}(UF)^{*}\begin{bmatrix} I_r \\ \mathbb{O} \end{bmatrix} = \begin{bmatrix} B, & K \end{bmatrix}(UF)^{*|r}$ *is invertible,*

(E_9) $\begin{bmatrix} I_r, & \mathbb{O} \end{bmatrix}(EV)^{*}\begin{bmatrix} B \\ K \end{bmatrix} = (EV)^{*}{}_{|r}\begin{bmatrix} B \\ K \end{bmatrix}$ *is invertible,*

(E_{10a}) $\begin{bmatrix} I_r, & T \end{bmatrix}(EF)^{*}\begin{bmatrix} I_r \\ S \end{bmatrix}$ *is invertible,*

(E_{10b}) $\begin{bmatrix} A_{11}, & A_{12} \end{bmatrix}(EF)^{*}\begin{bmatrix} A_{11} \\ A_{12} \end{bmatrix}$ *is invertible,*

(E_{11}) $\begin{bmatrix} I_r, & T_1^{-1}T_2 \end{bmatrix}\begin{bmatrix} T_1 \\ \mathbb{O} \end{bmatrix} = T_1$ *is invertible.*

Also, group inverse, if it exists, it has the following block representations:

$(A^1_{\#})$ $A^{\#} = R^{-1|r}\left((RG)^{-1}{}^{|r}_{|r}\right)^{-2}G^{-1}{}_{|r}$,

$(A^2_{\#})$ $A^{\#} = R^{-1|r}B^{-1}\left((RG)^{-1}{}^{|r}_{|r}\right)^{-2}G^{-1}{}_{|r}$,

$(A^3_{\#})$ $A^{\#} = R^{-1|r}\left((RF)^{-1}{}^{|r}_{|r} + K(RF)^{-1}{}^{|r}_{n-r|}\right)^{-2}\left(F^{*}{}_{|r} + KF^{*}_{n-r|}\right)$,

$(A^4_{\#})$ $A^{\#} = \left(E^{*|r} + E^{*n-r|}K\right)\left((EG)^{-1}{}^{|r}_{|r} + (EG)^{-1}{}^{n-r|}_{|r}K\right)^{-2}G^{-1}{}_{|r}$,

$(A^5_{\#})$ $A^{\#} = U^{*|r}\left((UG)^{-1}{}^{|r}_{|r}\right)^{-2}G^{-1}{}_{|r}$,

$(A^6_{\#})$ $A^{\#} = R^{-1|r}\left((RV)^{-1}{}^{|r}_{|r}\right)^{-2}V^{*}{}_{|r}$,

$(A^7_{\#})$ $A^{\#} = U^{*|r}B^{-1}\left((UV)^{*}{}^{|r}_{|r}\right)^{-2}V^{*}{}_{|r}$,

$(A^8_{\#})$ $A^{\#} = U^{*|r}\left(\begin{bmatrix} B, & K \end{bmatrix}(UF)^{*}\right)^{-2}\begin{bmatrix} B, & K \end{bmatrix}F^{*}$,

$(A^9_{\#})$ $A^{\#} = E^{*}\begin{bmatrix} B \\ K \end{bmatrix}\left((EV)^{*}{}_{|r}\begin{bmatrix} B \\ K \end{bmatrix}\right)^{-2}V^{*}{}_{|r}$,

$$(A_{\#}^{10a})\ A^{\#} = E^{*}\begin{bmatrix} I_r \\ S \end{bmatrix} A_{11}\left(\begin{bmatrix} I_r, & T \end{bmatrix}(EF)^{*}\begin{bmatrix} I_r \\ S \end{bmatrix} A_{11}\right)^{-2}\begin{bmatrix} I_r, & T \end{bmatrix} F^{*},$$

$$(A_{\#}^{10b})\ A^{\#} = E^{*}\begin{bmatrix} A_{11} \\ A_{21} \end{bmatrix} A_{11}\left(\begin{bmatrix} A_{11}, & A_{12} \end{bmatrix}(EF)^{*}\begin{bmatrix} A_{11} \\ A_{21} \end{bmatrix}\right)^{-2}\begin{bmatrix} A_{11}, & A_{12} \end{bmatrix} F^{*},$$

$$(A_{\#}^{11})\quad A^{\#} = R^{-1}\begin{bmatrix} T_1^{-1} & T_1^{-2}T_2 \\ \mathbb{O} & \mathbb{O} \end{bmatrix} R = R^{-1}{}_{|r}\begin{bmatrix} T_1^{-1} & T_1^{-2}T_2 \end{bmatrix} R.$$

Proof. Follows directly from the following: If the $A = PQ$ full-rank factorization of A, then $A^{\#}$ exists if and only if QP invertible matrix, and $A^{\#} = P(QP)^{-2}Q$. □

Methods for calculating the inverse group can be implemented on follows:

```
Group1[a_List]:=Block[{ran,r,g,rg,pom1,pom2},
   {r,g}=t1[a]; ran=rank[a];
   rg=r.g;
   rg=Inverse[rg];
   rg=Take[rg,ran];
   rg=Transpose[Take[Transpose[rg],ran]];
   If[rank[rg]==ran,
   pom1=Transpose[Take[Transpose[Inverse[r]],ran]];
   pom2=Take[Inverse[g],ran];
   Return[pom1.MatrixPower[Inverse[rg],2].pom2],
   Return[Group inverse does not exist"];
   ]; ]

Group3[a_List]:=Block[{ran,r,g,pom1,pom2},
   {r,f}=t3[a]; ran=rank[a];
   pom1=Transpose[Take[Transpose[Inverse[r]],ran]];
   pom2=Take[r.a.f,ran].Hermit[f];
   pom3=pom2.pom1;
   If[rank[pom3]==ran,
   Return[pom1.MatrixPower[Inverse[pom3],2].pom2],
   Return[Group inverse does not exist"];
   ]; ]
```

```
Group4[a_List]:=Block[{ran,r,g,k,pom1,pom2},
  {e,g}=t3[a]; ran=rank[a];
  pom1=Take[Inverse[g],ran];
  pom2=Hermit[e].Transpose[Take[Transpose[e.a.g],ran]];
  pom3=pom2.pom1;
  If[rank[pom3]==ran,
  Return[pom2.MatrixPower[Inverse[pom3],2].pom1],
  Return[Group inverse does not exist"];
  ]; ]
```

Example 3.11.4. For *PQ* factorization of matrix $A = \begin{bmatrix} 9 & 6 & 9 \\ 6 & 6 & 6 \\ 9 & 6 & 9 \end{bmatrix}$ we have $R = \begin{bmatrix} \frac{1}{9} & 0 & 0 \\ -\frac{1}{3} & \frac{1}{2} & 0 \\ -1 & 0 & 1 \end{bmatrix}$, $G = \begin{bmatrix} 1 & -\frac{2}{3} & -1 \\ 0 & 1 & 0 \\ 0 & 0 & 1 \end{bmatrix}$

$$R^{-1} = \begin{bmatrix} 9 & 0 & 0 \\ 6 & 2 & 0 \\ 9 & 0 & 1 \end{bmatrix}, G^{-1} = \begin{bmatrix} 1 & \frac{2}{3} & 1 \\ 0 & 1 & 0 \\ 0 & 0 & 1 \end{bmatrix}$$

$$P = R^{-1} \begin{bmatrix} I_2 \\ \mathbb{O} \end{bmatrix} = \begin{bmatrix} 9 & 0 & 0 \\ 6 & 2 & 0 \\ 9 & 0 & 1 \end{bmatrix} \begin{bmatrix} 1 & 0 \\ 0 & 1 \\ 0 & 0 \end{bmatrix} = \begin{bmatrix} 9 & 0 \\ 6 & 2 \\ 9 & 0 \end{bmatrix}$$

$$Q = \begin{bmatrix} I_2 & \mathbb{O} \end{bmatrix} G^{-1} = \begin{bmatrix} 1 & 0 & 0 \\ 0 & 1 & 0 \end{bmatrix} \begin{bmatrix} 1 & \frac{2}{3} & 1 \\ 0 & 1 & 0 \\ 0 & 0 & 1 \end{bmatrix} = \begin{bmatrix} 1 & \frac{2}{3} & 1 \\ 0 & 1 & 0 \end{bmatrix}$$

Example 3.11.5. Let be given the matrix

$$A = \begin{bmatrix} -1 & 1 & 3 & 5 & 7 \\ 1 & 0 & -2 & 0 & 4 \\ 1 & 1 & -1 & 5 & 15 \\ -1 & 2 & 4 & 10 & 18 \end{bmatrix}.$$

From $\{r1, f1\} = t3[a]$ we get

$$R_1 = \begin{bmatrix} 0 & 1 & 0 & 0 \\ 1 & 1 & 0 & 0 \\ -1 & -2 & 1 & 0 \\ -2 & -1 & 0 & 1 \end{bmatrix}, \quad F_1 = I_5$$

Applying $\{r2, f2\} = t3[Transpose[r1, a, f1]]$ we get

$$R_2 = \begin{bmatrix} 1 & 0 & 0 & 0 & 0 \\ 0 & 1 & 0 & 0 & 0 \\ 2 & -1 & 1 & 0 & 0 \\ 0 & -5 & 0 & 1 & 0 \\ -4 & -11 & 0 & 0 & 1 \end{bmatrix}, \quad F_2 = I_4.$$

Now we have $R = F_2^T R_1, G = F_1 R_2^T$, i

$$R_{|2}^{-1} = \begin{bmatrix} -1 & 1 \\ 1 & 0 \\ 1 & 1 \\ -1 & 2 \end{bmatrix}, \quad G_{|2}^{-1} = \begin{bmatrix} 1 & 0 & -2 & 0 & 4 \\ 0 & 1 & 1 & 5 & 11 \end{bmatrix}.$$

Using the formula $\{A_1^\dagger\}$ we get

$$A^\dagger = \begin{bmatrix} -\frac{169}{6720} & \frac{67}{2240} & \frac{233}{6720} & -\frac{137}{6720} \\ \frac{1}{128} & -\frac{1}{128} & -\frac{1}{128} & \frac{1}{128} \\ \frac{781}{13440} & -\frac{330}{4480} & -\frac{1037}{13440} & \frac{653}{13440} \\ \frac{5}{128} & -\frac{5}{128} & -\frac{5}{128} & \frac{5}{128} \\ -\frac{197}{13440} & \frac{151}{4480} & \frac{709}{13440} & \frac{59}{13440} \end{bmatrix}$$

Example 3.11.6. For $A = \begin{bmatrix} -1 & 0 & 1 & 2 \\ -1 & 1 & 0 & -1 \\ 0 & -1 & 1 & 3 \\ 0 & 1 & -1 & -3 \\ 1 & -1 & 0 & 1 \\ 1 & 0 & -1 & -2 \end{bmatrix}$ we get

$$A_{11}^{-1} = \begin{bmatrix} -1 & 0 \\ -1 & 1 \end{bmatrix}, \quad S = \begin{bmatrix} 1 & -1 \\ -1 & 1 \\ 0 & -1 \\ -1 & 0 \end{bmatrix}, \quad T = \begin{bmatrix} -1 & -2 \\ -1 & -3 \end{bmatrix}.$$

Applying the Theorem 2.9.11, and the Theorem 2.9.12 for

$$W_1 = \begin{bmatrix} 2 & 0 \\ 0 & 1 \\ 1 & 0 \\ 4 & 2 \end{bmatrix}, \quad W_2 = \begin{bmatrix} 3 & 1 & 3 & 1 & 2 & -1 \\ 0 & -1 & 0 & 0 & -2 & 1 \end{bmatrix}.$$

we get:

$$A^{(1,2,3)} = \begin{bmatrix} -\frac{2}{17} & -\frac{3}{17} & \frac{1}{17} & -\frac{1}{17} & \frac{3}{17} & \frac{2}{17} \\ \frac{19}{102} & \frac{10}{51} & -\frac{1}{102} & \frac{1}{102} & -\frac{10}{51} & -\frac{19}{102} \\ -\frac{1}{17} & -\frac{3}{34} & \frac{1}{34} & -\frac{1}{34} & \frac{3}{34} & \frac{1}{17} \\ \frac{7}{51} & \frac{2}{51} & \frac{5}{51} & -\frac{5}{51} & -\frac{2}{51} & -\frac{7}{51} \end{bmatrix},$$

$$A^{(1,2,4)} = \begin{bmatrix} -\frac{11}{17} & \frac{43}{51} & -\frac{11}{17} & -\frac{11}{51} & \frac{86}{51} & -\frac{43}{51} \\ \frac{7}{17} & -\frac{32}{51} & \frac{7}{17} & \frac{7}{51} & -\frac{64}{51} & \frac{32}{51} \\ \frac{4}{17} & -\frac{11}{51} & \frac{4}{17} & \frac{4}{51} & -\frac{22}{51} & \frac{11}{51} \\ \frac{1}{17} & \frac{10}{51} & \frac{1}{17} & \frac{1}{51} & \frac{20}{51} & -\frac{10}{51} \end{bmatrix},$$

$$A^{\dagger} = \begin{bmatrix} -\frac{5}{34} & -\frac{3}{17} & \frac{1}{34} & -\frac{1}{34} & \frac{3}{17} & \frac{5}{34} \\ \frac{4}{51} & \frac{13}{102} & -\frac{5}{102} & \frac{5}{102} & -\frac{13}{102} & -\frac{4}{51} \\ \frac{7}{102} & \frac{5}{102} & \frac{1}{51} & -\frac{1}{51} & -\frac{5}{102} & -\frac{7}{102} \\ \frac{1}{17} & -\frac{1}{34} & \frac{3}{34} & -\frac{3}{34} & \frac{1}{34} & -\frac{1}{17} \end{bmatrix}.$$

Example 3.11.7. Notice, by using the same matrix the extended Gauss-Jordan transformation, we obtain the following reduced row-echelon form for A:

$$B=\begin{bmatrix} I_r & K \\ \mathbb{O} & \mathbb{O} \end{bmatrix}=\begin{bmatrix} 1 & 0 & -1 & -2 \\ 0 & 1 & -1 & -3 \\ 0 & 0 & 0 & 0 \\ 0 & 0 & 0 & 0 \\ 0 & 0 & 0 & 0 \\ 0 & 0 & 0 & 0 \end{bmatrix}.$$

The matrix B is obtained by using the permutation matrix $E=I_4$, and regular matrix

$$R=\begin{bmatrix} -1 & 0 & 0 & 0 & 0 & 0 \\ -1 & 1 & 0 & 0 & 0 & 0 \\ -1 & 1 & 1 & 0 & 0 & 0 \\ 1 & -1 & 0 & 1 & 0 & 0 \\ 0 & 1 & 0 & 0 & -1 & 0 \\ 1 & 0 & 0 & 0 & 0 & 1 \end{bmatrix}.$$

Applying Theorem 1.3.1, Theorem 1.3.2, 1.3.1 and consequences of the same matrix W_1 i W_2 we get:

$$A^{(1,2)}=\begin{bmatrix} -\frac{2}{51} & -\frac{6}{17} & \frac{10}{17} & \frac{10}{51} & -\frac{14}{51} & \frac{8}{17} \\ \frac{23}{51} & \frac{1}{17} & -\frac{13}{17} & -\frac{13}{51} & \frac{8}{51} & -\frac{7}{17} \\ \frac{1}{51} & -\frac{3}{17} & \frac{5}{17} & \frac{5}{51} & -\frac{7}{51} & \frac{4}{17} \\ \frac{14}{17} & -\frac{10}{17} & -\frac{6}{17} & -\frac{2}{17} & -\frac{4}{17} & \frac{2}{17} \end{bmatrix},$$

$$A^{(1,2,4)}=\begin{bmatrix} -\frac{13}{51} & -\frac{3}{17} & \frac{11}{17} & \frac{11}{51} & -\frac{10}{51} & \frac{7}{17} \\ -\frac{1}{51} & \frac{5}{17} & -\frac{7}{17} & -\frac{7}{51} & \frac{11}{51} & -\frac{6}{17} \\ \frac{14}{51} & -\frac{2}{17} & -\frac{4}{17} & -\frac{4}{51} & -\frac{1}{51} & -\frac{1}{17} \\ \frac{29}{51} & -\frac{9}{17} & -\frac{1}{17} & -\frac{1}{51} & -\frac{13}{51} & \frac{4}{17} \end{bmatrix}.$$

$A^{\dagger}$ i $A^{(1,2,3)}$ are as in example 1.3.4.

Example 3.11.8. For matrix $R = \begin{bmatrix} -1 & 0 & 1 & 2 \\ -1 & 1 & 0 & -1 \\ 0 & -1 & 1 & 3 \\ 1 & 1 & -2 & -5 \end{bmatrix}$ we get

$$R = \begin{bmatrix} -1 & 0 & 0 & 0 \\ -1 & 1 & 0 & 0 \\ -1 & 1 & 1 & 0 \\ 2 & -1 & 0 & 1 \end{bmatrix}, \quad T_1 = \begin{bmatrix} 1 & 0 \\ 0 & 1 \end{bmatrix}, \quad T_2 = \begin{bmatrix} -1 & -2 \\ -1 & -3 \end{bmatrix}, \quad E = I_4.$$

Using matrices

$$W_1 = \begin{bmatrix} 2 & 0 \\ 0 & 1 \\ 1 & 0 \\ 4 & 2 \end{bmatrix}, \quad W_2 = \begin{bmatrix} 3 & 1 & 3 & 1 \\ 0 & -1 & 0 & 0 \end{bmatrix},$$

we get:

$$A^{(1,2,3)} = \begin{bmatrix} -\frac{10}{51} & -\frac{6}{17} & \frac{8}{51} & \frac{2}{51} \\ \frac{13}{51} & \frac{20}{51} & -\frac{7}{51} & -\frac{2}{17} \\ -\frac{5}{51} & -\frac{3}{17} & \frac{4}{51} & \frac{1}{51} \\ \frac{2}{17} & \frac{4}{51} & \frac{2}{51} & -\frac{8}{51} \end{bmatrix},$$

$$A^{(1,2,4)} = \begin{bmatrix} -\frac{120}{221} & -\frac{40}{221} & \frac{2}{13} & -\frac{1}{13} \\ \frac{81}{221} & \frac{27}{221} & -\frac{2}{13} & \frac{1}{13} \\ \frac{3}{17} & \frac{1}{17} & 0 & 0 \\ -\frac{3}{221} & -\frac{1}{221} & \frac{2}{13} & -\frac{1}{13} \end{bmatrix},$$

$$A^{\dagger} = \begin{bmatrix} -\frac{11}{51} & -\frac{6}{17} & \frac{7}{51} & \frac{4}{51} \\ \frac{7}{51} & \frac{13}{51} & -\frac{2}{17} & -\frac{1}{51} \\ \frac{4}{51} & \frac{5}{51} & -\frac{1}{51} & -\frac{1}{17} \\ \frac{1}{51} & -\frac{1}{17} & \frac{4}{51} & -\frac{5}{51} \end{bmatrix},$$

$$A^{\#} = \begin{bmatrix} -5 & 4 & 1 & -2 \\ -21 & 17 & 4 & -9 \\ 16 & -13 & -3 & 7 \\ -11 & 9 & 2 & -5 \end{bmatrix}.$$

Corollary 3.11.16. *Let $A \in \mathbb{C}_r^{m\times n}$ and T,S subsets of $\mathbb{C}^n$ and $\mathbb{C}^m$, respectively, such that $dimT = dimS^{\perp} = t \leq r$ i $AT \oplus S = \mathbb{C}^m$. Then exists $\{2\}$-inverse of $A, X = A^{(2)}_{T,S}$, such that $R(X) = T, N(X) = S$. Moreover, if $D \in \mathbb{C}^{n\times m}, C \in \mathbb{C}^{n\times t}$ satisfies $R(D) = T, N(D) = S, R(C) = T$, Then bloc presentation for $A^{(2)}_{T,S}$ can be exercised in the same way as a block representation $(G_2^i), i = 1,\ldots,n$ using the following shifts:*

$$W_1 = C,\ W_2 = C^*D.$$

Theorem 3.11.17. *Let $A \in \mathbb{C}_r^{n\times n}$ satisfies $k = indA$ and the following expression block decomposition form T_{11}:*

$$A_i = R_i^{-1} \begin{bmatrix} A_{i+1} & B_{i+1} \\ \mathbb{O} & \mathbb{O} \end{bmatrix} R_i, i = 1,\ldots,k,$$

where $A_1 = A, r_1 = r$ and $A_i + 1$ the rows of square blocks r_1. Then Drazin's inverse of A can be calculated as follows:

$$A^D \begin{cases} \left\{\prod_{i=1}^{k} P_i \left(\prod_{j=k}^{1} Q_j A \prod_{i=1}^{k} P_i\right)^{-1} \prod_{j=k}^{1} Q_j & Q_k P_k \text{ is invertible} \\ \mathbb{O} & Q_k P_k = \mathbb{O}. \end{cases} \tag{3.11.2}$$

where

$$P_i = R_i^{-1} \begin{bmatrix} I_r \\ \mathbb{O} \end{bmatrix} = (R_i^{-1})_i^r,$$

$$Q_i = \begin{bmatrix} A_{i+1}, & B_{i+1} \end{bmatrix} R_i, \quad i = 1\ldots,k.$$

3.12 Block LDL* Decomposition of the Full-Rank

The Cholesky decomposition is often used to calculate the inverse matrix A^{-1} and the determinant of A. For a given Hermitian positive definite matrix A (see [3]), there exists a nonsingular lower triangular matrix L such that

$$A = LL^*.$$

If A is not a positive definite matrix, the matrix L can be singular as well. One can also impose that the diagonal elements of the matrix L are all positive, in which case the corresponding factorization is unique. The difficulty involving the Cholesky decomposition is the appearance of square root entries in matrix L, which is sometimes called the "square" of A.

Procedure to determine the block Cholesky decomposition includes calculating the square root of the diagonal block and then multiplication inverse triangular block and a square block. Determining a square root block matrix can be difficult and long task. Several methods are known for finding approximations, such as block LU and block Cholesky factorization (see [8]). In doing so, multiplication inverse triangular block with block matrix can be achieved by so-called "Back substitution" process.

Replacement of LL* decomposition by LDL* to avoid elements with square roots is well explained techniques from linear algebra (see for example the work of Golub and Van Loan [19]). LDL* factorization requires less calculation of the Gaussian Elimination. It is generated by an additional diagonal matrix D, or a total amount of calculation is the same as in the Cholesky decomposition. Note that even for a given Hermitian matrix A, D matrix must have a non-zero positive elements.

Consider an arbitrary set of normal matrix equations. It can be expressed in the following matrix form: $A\mathrm{x} = B$, where x is n-dimensional vector solutions. Here are some preliminary results from the work [57] to calculate the vector x, based on LDL^* factorization of the matrix A. Let the matrix A be Hermitian and positive definite. Then, there are nonsingular lower triangular matrix L and nonsingular diagonal matrix D such that

$$A = LDL^*.$$

Inverse matrix A^{-1} and the vector of solutions x can be determined in the following ways:

$$\begin{aligned} A^{-1} &= L^{*-1} \cdot D^{-1} \cdot L^{-1}, \\ \mathrm{x} &= L^{*-1} \cdot D^{-1} \cdot L^{-1} \cdot B. \end{aligned} \tag{3.12.1}$$

These equations include inverse triangular and diagonal matrix. However, it is not necessary, because it is possible to grant triangular matrix inverse matrix without calculating. Therefore, consider the next move:

$$C = DL^*\mathrm{x}.$$

Then $L \cdot C = B$ and the following is valid:

$$c_i = \frac{b_i - \sum_{k=1}^{i-1} l_{i,k} \cdot c_k}{l_{i,i}}, \quad for\, i = \overline{1,n}. \tag{3.12.2}$$

It is obvious that the product matrix DL^* upper triangular matrix, and therefore the following expressions are valid

$$x_n = \frac{c_n}{d_{n,n}}, \tag{3.12.3}$$

$$x_i = \frac{c_i - \sum_{k=i+1}^{n} d_{i,i} l_{k,i} \cdot x_k}{d_{i,i}}, \quad for\, i = \overline{n-1,1} \tag{3.12.4}$$

The main motivation is to determine the procedure for the block LDL* decomposition, the general case $n \times n$ matrix block Hermitian $A \in \mathbf{C}^{m \times m}$. This is to avoid difficulties in calculating roots elements of sub matrices (often called blocks).

Let the Hermitian matrix A is divided into $n \times n$ blocks, so that $A = [A_{ij}]$, $1 \leq i, j \leq n$, whereby the diagonal of a square matrix blocks. It is necessary to determine the lower block triangular matrix L and block diagonal matrix D, such that the equality $A = LDL^*$ is satisfied, with the dimensions of the sub-matrices $L_{i,j}$ and $D_{i,j}$ are equal to $d_i \times d_j$. A simple modification of the iterative procedure for LDL^* decomposition, which applies to block elements of the matrices D and L is given in Ref. [57].

Theorem 3.12.1. [57] *Let the data block Hermitian matrix*$A = [A_{ij}]_{n\times n}$ *rank* r, *divided in such a way that dimension of each sub-matrix* $A_{i,j}$ *is equal* $d_i \times d_j$, *and that there exists a natural number* $m \leq n$ *then the following expression is valid* $\sum_{k=1}^{m} d_k = r$. *Let* $L = [L_{ij}]_{n\times m}$ *i* $D = [D_{ij}]_{m\times m}$ *block matrix, where dimension of each sub-matrix* L_{ij} *i* D_{ij} *is equal* $d_i \times d_j$. *If the vase next to the matrix equations* $j = \overline{1,m}$:

$$D_{jj} = A_{jj} - \sum_{k=1}^{j-1} L_{jk} \cdot D_{kk} \cdot L_{jk}^*, \tag{3.12.5}$$

$$L_{ij} = (A_{ij} - \sum_{k=1}^{j-1} L_{ik} \cdot D_{kk} \cdot L_{jk}^*) \cdot D_{jj}^{-1}, \quad i = \overline{j+1,n}, \tag{3.12.6}$$

whereby the sub-matrix D_{jj}, $j = \overline{1,m}$ *are nonsingular, then* LDL^* *is full-rank decomposition of matrix* A.

Proof. How $A = LDL^*$, to the an arbitrary indices $1 \leq i,j \leq n$ the following identity is valid:

$$A_{ij} = \sum_{k=1}^{m} L_{ik} D_{kk} L_{jk}^* = \sum_{k=1}^{\min\{i,j\}} L_{ik} D_{kk} L_{jk}^*,$$

given that following the matrix equality met for each $i = \overline{1,m}$:

$$L_{ii} = \mathbf{I}_{d_i},$$
$$L_{ij} = \mathbf{0}_{d_i \times d_j}, \; D_{ij} = \mathbf{0}_{d_i \times d_j}, \; D_{ji} = \mathbf{0}_{d_j \times d_i}, \quad j = \overline{i+1,m}.$$

Therefore, for each index $j = \overline{1,m}$:

$$A_{jj} = \sum_{k=1}^{j} L_{jk} D_{kk} L_{jk}^* = D_{jj} + \sum_{k=1}^{j-1} L_{jk} D_{kk} L_{jk}^*$$

the equality is valid, from which follows (3.12.5). Therefore, for case where $i > j$ we have

$$A_{ij} = \sum_{k=1}^{j} L_{ik} D_{kk} L_{jk}^* = L_{ij} D_{jj} L_{jj}^* + \sum_{k=1}^{j-1} L_{ik} D_{kk} L_{jk}^*.$$

So satisfied is as follows:

$$L_{ij} D_{jj} = A_{ij} - \sum_{k=1}^{j-1} L_{ik} D_{kk} L_{jk}^*,$$

and thus the equation (3.12.6) is valid, because the submatrix D_{jj} is non-singular.

Given that the term $\sum_{k=1}^{m} d_k = r$ is satisfied, matrices $L = [L_{ij}]_{n\times m}$ and $D = [D_{ij}]_{m\times m}$ are both rank r, based on equality (3.12.5) i (3.12.6). Thus, LDL^* is full-rank decomposition of matrix A. □

Obviously, iterative procedure to block LDL* Hermitian matrix decomposition is can be determined from equations (3.12.5) and (3.12.6). In each iteration, the sub-matrix D_{jj} and L_{ij}, $i > j$ can be calculated on the basis of previous results and known blocks A_{ij}.

Note that the requirement of sub-matrix D_{jj}, $j = \overline{1,n}$ are non-singular may be too high in some cases teas. These conditions can be avoided using the Moore–Penrose's inverse matrix's D_{jj}, in which case the equation (3.12.6) station:

$$L_{ij} = (A_{ij} - \sum_{k=1}^{j-1} L_{ik} \cdot D_{kk} \cdot L_{jk}^*) \cdot D_{jj}^\dagger, \quad i = \overline{j+1,n}. \tag{3.12.7}$$

Iterative procedures for LDL* decomposition solution block vector is similar as in this approach.

So, as an extension of Theorem 3.12.1, the following algorithm, which can generate the full-rank of block matrix $L\,D$ dimensions smaller than or equal to the dimensions of the matrix A, can be formulated.

Algorithm 3.11 Block LDL^* full-rank decomposition of divided block matrix

Input: Block matrix $A = [A_{ij}]_{i,j=1}^{n}$ rank r, where the dimensions of an arbitrary block $A_{i,j}$ is equal to $d_i \times d_j$, and for any $m \in \{1,\ldots,n\}$ is satisfied identity $\sum\limits_{k=1}^{m} d_k = r$.

1: **Initialization**: for $i = \overline{1,m}$ perform the following steps:

1.1: Set $L_{ii} = \mathbf{I}_{d_i}$.

1.2: For $j = \overline{i+1,n}$ set $L_{ij} = \mathbf{0}_{d_i \times d_j}$.

1.2: For $j = \overline{i+1,m}$ set $D_{ij} = \mathbf{0}_{d_i \times d_j}$, $D_{ji} = \mathbf{0}_{d_j \times d_i}$.

2: **Evaluation**: for each $j = \overline{1,m}$ perform steps 2.1 and 2.2:

2.1: Set $D_{jj} = A_{jj} - \sum_{k=1}^{j-1} L_{jk} \cdot D_{kk} \cdot L_{jk}^*$.

2.2: For $i = \overline{j+1,n}$ sat $L_{ij} = (A_{ij} - \sum_{k=1}^{j-1} L_{ik} \cdot D_{kk} \cdot L_{jk}^*) \cdot D_{jj}^{-1}$.

Example 3.12.1. Consider the following 3×3 the block matrix A:

$$A = \left[\begin{array}{ccc|cc|c} 5 & 6 & 7 & 5 & 6 & 7 \\ 6 & 9 & 10 & -6 & 9 & 10 \\ 7 & 10 & 13 & 7 & 10 & 13 \\ \hline 5 & -6 & 7 & 5 & -6 & -7 \\ 6 & 9 & 10 & -6 & -9 & 12 \\ \hline 7 & 10 & 13 & -7 & 12 & -13 \end{array}\right].$$

After initialization sub-matrix $D_{12}, D_{13}, D_{21}, D_{23}, D_{31}, D_{32}, L_{12}, L_{13}, L_{23}$, Execute Step 2 in Algorithm 3.11. For $j = 1$, sub-matrix D_{11} is equal A_{11}, and is therefore, in Step 2.2:

$$\begin{aligned} L_{21} &= A_{21} \cdot D_{11}^{-1} = \begin{bmatrix} 7 & -12 & 6 \\ 0 & 1 & 0 \end{bmatrix}, \\ L_{31} &= A_{31} \cdot D_{11}^{-1} = \begin{bmatrix} 0 & 0 & 1 \end{bmatrix}. \end{aligned}$$

In the second iteration, for $j = 2$ calculated by sub-matrices D_{22} and L_{32}:

$$\begin{aligned} D_{22} &= A_{22} - L_{21}D_{11}L_{21}^* = \begin{bmatrix} -144 & 0 \\ 0 & -18 \end{bmatrix}, \\ L_{32} &= (A_{32} - L_{31} \cdot D_{11} \cdot L_{21}^*) \cdot D_{22}^{-1} = \begin{bmatrix} \frac{7}{72} & \frac{11}{9} \end{bmatrix}. \end{aligned}$$

Finally, in case $j = 3$, we have that

$$D_{33} = A_{33} - L_{31} \cdot D_{11} \cdot L_{31}^* - L_{32} \cdot D_{22} \cdot L_{32}^* = \begin{bmatrix} \frac{9}{4} \end{bmatrix},$$

However, the $A = LDL^*$ block decomposition of the full-rank of the matrix A, where

$$L = \left[\begin{array}{ccc|cc|c} 1 & 0 & 0 & 0 & 0 & 0 \\ 0 & 1 & 0 & 0 & 0 & 0 \\ 0 & 0 & 1 & 0 & 0 & 0 \\ \hline 7 & -12 & 6 & 1 & 0 & 0 \\ 0 & 1 & 0 & 0 & 1 & 0 \\ \hline 0 & 0 & 1 & \frac{7}{72} & \frac{11}{9} & 1 \end{array}\right],$$

$$D = \left[\begin{array}{ccc|cc|c} 5 & 6 & 7 & 0 & 0 & 0 \\ 6 & 9 & 10 & 0 & 0 & 0 \\ 7 & 10 & 13 & 0 & 0 & 0 \\ \hline 0 & 0 & 0 & -144 & 0 & 0 \\ 0 & 0 & 0 & 0 & -18 & 0 \\ \hline 0 & 0 & 0 & 0 & 0 & \frac{9}{4} \end{array}\right]$$

It is obvious that the total number of non-zero elements in the matrix L and D decreases the further splitting of the matrix A. Matrix L and D in some point of sharing can be considered rare.

Example 3.12.2. The main difference between the bloc LDL^* decomposition of the full-rank of standard and its variants is visible in the class rank-deficient matrix. In this sense, observe that A_2 is matrix of rank $r = 2$

from Ref. [88], divided as follows:

$$A_2 = \left[\begin{array}{cc|cc|cc} a+10 & a+9 & a+8 & a+7 & a+6 & a+5 \\ a+9 & a+8 & a+7 & a+6 & a+5 & a+4 \\ \hline a+8 & a+7 & a+6 & a+5 & a+4 & a+3 \\ a+7 & a+6 & a+5 & a+4 & a+3 & a+2 \\ \hline a+6 & a+5 & a+4 & a+3 & a+2 & a+1 \\ a+5 & a+4 & a+3 & a+2 & a+1 & a \end{array}\right]$$

Applying Algorithm 3.11 following matrix full-rank were obtained as a result for $m = 1$:

$$L = \left[\begin{array}{cc} 1 & 0 \\ 0 & 1 \\ \hline -1 & 2 \\ -2 & 3 \\ \hline -3 & 4 \\ -4 & 5 \end{array}\right], \quad D = \left[\begin{array}{cc} 10+a & 9+a \\ 9+a & 8+a \end{array}\right].$$

Example 3.12.3. Consider the rank-deficient matrix symbolic F_6 obtained in the Ref. [88], divided as follows:

$$F_6 = \left[\begin{array}{cc|ccc|c} a+6 & a+5 & a+4 & a+3 & a+2 & a+1 \\ a+5 & a+5 & a+4 & a+3 & a+2 & a+1 \\ \hline a+4 & a+4 & a+4 & a+3 & a+2 & a+1 \\ a+3 & a+3 & a+3 & a+3 & a+2 & a+1 \\ a+2 & a+2 & a+2 & a+2 & a+1 & a \\ \hline a+1 & a+1 & a+1 & a+1 & a & a-1 \end{array}\right].$$

Notice that $\text{rank}(F_6) = 5$, and because of that block matrices LDL^* full-rank factorizations of the following matrix $L \in \mathbb{C}(a)_5^{6\times 5}$ and $D \in \mathbb{C}(a)_5^{5\times 5}$. Applying the Algorithm 3.11 on the matrix F_6, for case $n = 3$, $m = 2$, generate the following rational matrix as a result of:

$$L=\left[\begin{array}{cc|ccc} 1 & 0 & 0 & 0 & 0 \\ 0 & 1 & 0 & 0 & 0 \\ \hline 0 & \frac{4+a}{5+a} & 1 & 0 & 0 \\ 0 & \frac{3+a}{5+a} & 0 & 1 & 0 \\ 0 & \frac{2+a}{5+a} & 0 & 0 & 1 \\ \hline 0 & \frac{1+a}{5+a} & 0 & -1 & 2 \end{array}\right], \quad D=\left[\begin{array}{cc|ccc} a+6 & a+5 & 0 & 0 & 0 \\ a+5 & a+5 & 0 & 0 & 0 \\ \hline 0 & 0 & \frac{4+a}{5+a} & \frac{3+a}{5+a} & \frac{2+a}{5+a} \\ 0 & 0 & \frac{3+a}{5+a} & \frac{2(3+a)}{5+a} & \frac{2(2+a)}{5+a} \\ 0 & 0 & \frac{2+a}{5+a} & \frac{2(2+a)}{5+a} & \frac{1+2a}{5+a} \end{array}\right].$$

3.12.1 Inverses $n \times n$ Matrix of Block

Matrix L and D, included in the bloc LDL^* decomposition, the lower block triangular and block diagonal matrix, respectively. Let's examine the first of these types of inverses of block matrices.

Lemma 3.12.2. *Let the data block diagonal matrix D with n sub-matrices on the diagonal. Then D invertible matrix if and only if D_i square and nonsingular matrix for every $1 \leq i \leq n$, in which case the D^{-1} also block diagonal matrix form*

$$D^{-1}=\begin{bmatrix} D_1^{-1} & 0 & \cdots & 0 \\ 0 & D_2^{-1} & \cdots & 0 \\ \vdots & \vdots & \ddots & \vdots \\ 0 & 0 & \cdots & D_n^{-1} \end{bmatrix}.$$

Lemma 3.12.3. [57] *Inverse nonsingular lower triangular matrices block*

$$L=\begin{bmatrix} I & 0 & \cdots & 0 \\ L_{21} & I & \cdots & 0 \\ \vdots & \vdots & \ddots & \vdots \\ L_{n1} & L_{n2} & \cdots & I \end{bmatrix},$$

the lower block triangular matrix form

$$L^{-1} = \begin{bmatrix} I & 0 & \cdots & 0 \\ (L^{-1})_{21} & I & \cdots & 0 \\ \vdots & \vdots & \ddots & \vdots \\ (L^{-1})_{n1} & (L^{-1})_{n2} & \cdots & I \end{bmatrix},$$

wherein $(L^{-1})_{i+1,i} = -L_{i+1,i}$ *for each index* $i = \overline{1,n-1}$.

Proof. Obviously, matrix L^{-1}, the lower block diagonal matrix with the identity matrix on the diagonal.

On the basis of equation $LL^{-1} = I$, we have a system of n^2 matrix equations. For an arbitrary $1 \le i < n$, by multiplying $(i+1)$-th row of matrix L as i-th column of inverse L^{-1} we have

$$L_{i+1,i}I + I(L^{-1})_{i+1,i} = 0,$$

so based on this we have $(L^{-1})_{i+1,i} = -L_{i+1,i}$. □

As we've established, LDL* decomposition of the matrix A can be used to determine its inverse:

$$A^{-1} = (L^{-1})^* \cdot D^{-1} \cdot L^{-1}. \quad (3.12.8)$$

Methods for the calculation of the inverse matrix Hermitian block LDL^* decomposition is based on this equation. So, from this it follows next theorem which is introduced in the Ref. [57].

Theorem 3.12.4. [57] *Consider the block* LDL^* *decomposition of nonsingular Hermitian block matrix* $A = [A_{ij}]_{i,j=1}^n$, *where the matrices* L *and* D *have the same division. If the matrix* L *and sub-matrices* D_{jj}, $j = \overline{1,n}$ *are nonsingular, then the following matrix equation is valid:*

$$(A^{-1})_{ij} = \sum_{k=j}^{n} (L^{-1})^*_{ki} D^{-1}_{kk} (L^{-1})_{kj}, \quad 1 \le i \le j \le n. \quad (3.12.9)$$

Proof. Follows from Theorems 3.12.4 and 3.12.3, it is given that they are valid the following identity:

$$A = \begin{bmatrix} A_{11} & A_{12} \\ A_{12}^* & A_{22} \end{bmatrix}.$$

□

Corollary 3.12.5. *Consider the block LDL^* decomposition of the matrix A, where the matrix L and D have the same division as the A. If the sub-matrix D_{11} D_{22} nonsingular, then the inverse A^{-1} has the following form:*

$$A^{-1} = \begin{bmatrix} D_{11}^{-1} + L_{21}^* D_{22}^{-1} L_{21} & -L_{21}^* D_{22}^{-1} \\ -D_{22}^{-1} L_{21} & D_{22}^{-1} \end{bmatrix}. \tag{3.12.10}$$

Proof. Follows from Theorems 3.12.4 and 3.12.3, it is given that they are valid the following identity:

$$\begin{aligned} A^{-1} &= \begin{bmatrix} L_{11}^* D_{11}^{-1} L_{11} + L_{21}^* D_{22}^{-1} L_{21} & -L_{21}^* D_{22}^{-1} L_{22} \\ -L_{22}^* D_{22}^{-1} L_{21} & L_{22}^* D_{22}^{-1} L_{22} \end{bmatrix} \\ &= \begin{bmatrix} D_{11}^{-1} + L_{21}^* D_{22}^{-1} L_{21} & -L_{21}^* D_{22}^{-1} \\ -D_{22}^{-1} L_{21} & D_{22}^{-1} \end{bmatrix}. \end{aligned} \tag{3.12.11}$$

□

Obviously this representation inverse of 2×2 block matrix is simpler than her similar representations from the work [34] which is based on the complement of Schur. In fact, the representation (3.12.10) is the result of Theorem 2.1 from Ref. [34], in case $D_{11} = A_{11}$, $L_{21} = A_{12}^* D_{11}^{-1}$ and Schur complement $D_{22} = A_{22} - A_{12}^* D_{11}^{-1} A_{12}$. Also, the conditions that D_{11} D_{22} nonsingular matrices are equivalent conditions that the matrix A_{11} and Schur complement of $A_{22} - A_{12}^* A_{11}^{-1} A_{12}$ nonsingular.

Example 3.12.4. Notice, the next split matrix:

$$A=\left[\begin{array}{cc|ccc} -1 & 6 & 5 & 6 & 7 \\ 6 & 9 & -6 & 0 & 10 \\ \hline 5 & -6 & 5 & -6 & -7 \\ 6 & 0 & -6 & -9 & 12 \\ 7 & 10 & -7 & 12 & -13 \end{array}\right].$$

Then the matrix L and D bloc from LDL^* factorization of the matrix A equal to:

$$L=\left[\begin{array}{cc|ccc} 1 & 0 & 0 & 0 & 0 \\ 0 & 1 & 0 & 0 & 0 \\ \hline -\frac{9}{5} & \frac{8}{15} & 1 & 0 & 0 \\ -\frac{6}{5} & \frac{4}{5} & 0 & 1 & 0 \\ -\frac{1}{15} & \frac{52}{45} & 0 & 0 & 1 \end{array}\right], \quad D=\left[\begin{array}{cc|ccc} -1 & 6 & 0 & 0 & 0 \\ 6 & 9 & 0 & 0 & 0 \\ \hline 0 & 0 & \frac{86}{5} & \frac{24}{5} & \frac{4}{15} \\ 0 & 0 & \frac{24}{5} & -\frac{9}{5} & \frac{62}{5} \\ 0 & 0 & \frac{4}{15} & \frac{62}{5} & -\frac{1084}{45} \end{array}\right].$$

To calculate A^{-1}, we have the following sequence of expressions:

$$\begin{aligned}
(A^{-1})_{11} &= \begin{bmatrix} -1 & 6 \\ 6 & 9 \end{bmatrix}^{-1} + \begin{bmatrix} -\frac{9}{5} & \frac{8}{15} \\ -\frac{6}{5} & \frac{4}{5} \\ -\frac{1}{15} & \frac{52}{45} \end{bmatrix}^T \cdot \begin{bmatrix} \frac{86}{5} & \frac{24}{5} & \frac{4}{15} \\ \frac{24}{5} & -\frac{9}{5} & \frac{62}{5} \\ \frac{4}{15} & \frac{62}{5} & -\frac{1084}{45} \end{bmatrix}^{-1} \cdot \begin{bmatrix} -\frac{9}{5} & \frac{8}{15} \\ -\frac{6}{5} & \frac{4}{5} \\ -\frac{1}{15} & \frac{52}{45} \end{bmatrix} \\
&= \begin{bmatrix} \frac{99}{656} & -\frac{33}{164} \\ -\frac{33}{164} & \frac{96}{205} \end{bmatrix}, \\
(A^{-1})_{12} &= (A^{-1})^*_{21} = -\begin{bmatrix} -\frac{9}{5} & \frac{8}{15} \\ -\frac{6}{5} & \frac{4}{5} \\ -\frac{1}{15} & \frac{52}{45} \end{bmatrix}^T \cdot \begin{bmatrix} \frac{86}{5} & \frac{24}{5} & \frac{4}{15} \\ \frac{24}{5} & -\frac{9}{5} & \frac{62}{5} \\ \frac{4}{15} & \frac{62}{5} & -\frac{1084}{45} \end{bmatrix}^{-1} \\
&= \begin{bmatrix} \frac{13}{328} & \frac{223}{984} & \frac{75}{656} \\ \frac{33}{410} & -\frac{481}{1230} & -\frac{25}{164} \end{bmatrix}, \\
(A^{-1})_{22} &= \begin{bmatrix} \frac{86}{5} & \frac{24}{5} & \frac{4}{15} \\ \frac{24}{5} & -\frac{9}{5} & \frac{62}{5} \\ \frac{4}{15} & \frac{62}{5} & -\frac{1084}{45} \end{bmatrix}^{-1} = \begin{bmatrix} \frac{69}{820} & -\frac{223}{2460} & -\frac{15}{328} \\ -\frac{223}{2460} & \frac{259}{820} & \frac{53}{328} \\ -\frac{15}{328} & \frac{53}{328} & \frac{27}{656} \end{bmatrix},
\end{aligned}$$

on the basis of which is the inverse of A equals

$$A^{-1} = \left[\begin{array}{cc|ccc} \frac{99}{656} & -\frac{33}{164} & \frac{13}{328} & \frac{223}{984} & \frac{75}{656} \\ -\frac{33}{164} & \frac{96}{205} & \frac{33}{410} & -\frac{481}{1230} & -\frac{25}{164} \\ \hline \frac{13}{328} & \frac{33}{410} & \frac{69}{820} & -\frac{223}{2460} & -\frac{15}{328} \\ \frac{223}{984} & -\frac{481}{1230} & -\frac{223}{2460} & \frac{259}{820} & \frac{53}{328} \\ \frac{75}{656} & -\frac{25}{164} & -\frac{15}{328} & \frac{53}{328} & \frac{27}{656} \end{array}\right].$$

3.12.2 Moore–Penrose's Inverses of 2×2 Blocks of Matrices

Inverses of 2×2 block matrices are well investigated (see, e.g., [34, 41]), and often appear in many equalization calculations.

Some explicit inversion formulas 2×2 block matrix

$$\begin{bmatrix} A & B \\ C & D \end{bmatrix}^{-1}$$

were introduced in the work [34], and are based on Schur complement $D - CA^{-1}B$ of matrix A.

Lemma 3.12.6. *Let be given 2×2 divided nonsingular matrix*

$$M = \begin{bmatrix} A & B \\ C & D \end{bmatrix},$$

wherein $A \in \mathbf{C}^{k \times k}$, $B \in \mathbf{C}^{k \times l}$, $C \in \mathbf{C}^{l \times k}$ and $D \in \mathbf{C}^{l \times l}$. If the sub matrices A is nonsingular, then Schur complement $S_A = D - CA^{-1}B$ matrix A in M also nonsingular, the inverse of a matrix M has the following form

$$M^{-1} = \begin{bmatrix} A^{-1} + A^{-1}BS_A^{-1}CA^{-1} & -A^{-1}BS_A^{-1} \\ -S_A^{-1}CA^{-1} & S_A^{-1} \end{bmatrix}. \qquad (3.12.12)$$

Many results specify the methods for calculating the Moore–Penrose's inverse of 2×2 block matrices (see, e.g., [26]). Also, some of the representations of inverse Drazin of 2×2 block matrices are obtained in Ref.

[13]. However, the general case $n \times n$ matrix of block, $n > 2$ is not as described, see, Ref. [57]. We perform methods for $n \times n$ matrix block and then watched special case of 2×2 2 block matrix.

Assertion analogous to Theorem 3.12.4 can be introduce to calculating the Moore–Penrose's inverse Hermitian $n \times n$ matrix block, watching its pads LDL^* decomposition of the full-rank.

Theorem 3.12.7. [57] *Let the data block Hermitian matrix $A = [A_{i,j}]_{n\times n}$ rank r, where dimensions are an arbitrary matrices $A_{i,j}$ is given with $d_i \times d_j$, and that there exists an integer $m \leq n$ such that a satisfied expression $\sum_{k=1}^{m} d_k = r$. Suppose that LDL^* block decomposition of the full-rank of the matrix A, where the sub-matrices D_{jj} $j = \overline{1,m}$ nonsingular. Then, the following matrix equation:*

$$(A^\dagger)_{ij} = \sum_{k=j}^{m} (L^\dagger)^*_{ki} D^{-1}_{kk} (L^\dagger)_{kj}, \quad 1 \leq i \leq j \leq n. \qquad (3.12.13)$$

Several methods for evaluation of the generalized inverses of a rational matrix were introduced in Ref. [58]. Based on the LDL^* decomposition of a corresponding matrix products, they provide an efficient way of calculating generalized inverses of constant matrices. The following theorem from Ref. [58] gives the practical expression to compute Moore–Penrose's inverse of a rational matrix.

Theorem 3.12.8. [58] *Consider the rational matrix $A \in \mathbf{C}(x)^{m\times n}_r$. If LDL^* is the full-rank decomposition of matrix $(A^*A)^2$, where $L \in \mathbf{C}(x)^{m\times r}$ and $D \in \mathbf{C}(x)^{r\times r}$, then it is satisfied:*

$$A^\dagger = L(L^*LDL^*L)^{-1}L^*(A^*A)^*A^*. \qquad (3.12.14)$$

If LDL^* is the full-rank decomposition of the matrix $(AA^*)^2$, $L \in \mathbf{C}(x)^{m\times r}$ and $D \in \mathbf{C}(x)^{r\times r}$, then it is satisfied:

$$A^\dagger = A^*(AA^*)^*L(L^*LDL^*L)^{-1}L^*. \qquad (3.12.15)$$

We consider the corollary of this theorem considering only constant complex matrices.

The following lemma from Ref. [74] was proved to be a simple extension of some well-known results. It derives a criteria of evaluating the inverse of a partitioned matrix, under a few assumptions.

Lemma 3.12.9. *Let $A(x)$ be nonsingular partitioned matrix, and let submatrix $A_{11}(x)$ be also nonsingular. Then*

$$A^{-1}(x) = \begin{bmatrix} A_{11}(x) & A_{12}(x) \\ A_{21}(x) & A_{22}(x) \end{bmatrix}^{-1} = \begin{bmatrix} B_{11}(x) & B_{12}(x) \\ B_{21}(x) & B_{22}(x) \end{bmatrix}$$

where

$$\begin{aligned} B_{11}(x) =& A_{11}(x)^{-1} + A_{11}(x)^{-1}A_{12}(x)B_{22}(x)A_{21}(x)A_{11}(x)^{-1} \\ B_{12}(x) =& -A_{11}(x)^{-1}A_{12}(x)B_{22}(x) \\ B_{21}(x) =& -B_{22}(x)A_{21}(x)A_{11}(x)^{-1} \\ B_{22}(x) =& (A_{22}(x) - A_{21}(x)A_{11}(x)^{-1}A_{12}(x))^{-1}. \end{aligned} \tag{3.12.16}$$

Now we are able to develop a method for calculating the Moore–Penrose's inverse of 2×2 block matrix, given in the following consequence of Theorems 3.12.7 and Lemma 3.12.9.

Theorem 3.12.10. [57] *Let $A\in\mathbf{C}^{m\times n}$ i $(A^*A)^2\in\mathbf{C}^{n\times n}$ 2×2 be block matrix in the following form*

$$A = \begin{bmatrix} A_{11} & A_{12} \\ A_{21} & A_{22} \end{bmatrix}, \quad (A^*A)^2 = \begin{bmatrix} B & C \\ C^* & E \end{bmatrix},$$

with the same division. Let's get the next shift, in case nonsingular matrix B:

$$\begin{aligned} X_{11} &= B + N^* + N + M^*EM, \\ X_{12} &= C + ME, \\ X_{22} &= E, \end{aligned} \tag{3.12.17}$$

where the matrix M, N are determined as

$$\begin{aligned} M &= C^*B^{-1}, \\ N &= CM. \end{aligned} \tag{3.12.18}$$

If matrix $X = \begin{bmatrix} X_{11} & X_{12} \\ X_{12}^* & X_{22} \end{bmatrix}$ *and sub-matrix* X_{11}, B *are nonsingular, then Moore–Penrose's inverse of matrix A are given on the following way:*

$$A^\dagger = \begin{bmatrix} Y_{11}\Sigma_{11} + (Y_{11}M^* + Y_{12})\Sigma_{21} \\ (MY_{11} + Y_{21})(\Sigma_{11} + M^*\Sigma_{21}) + (MY_{12} + Y_{22})\Sigma_{21} \end{bmatrix.$$
$$\begin{matrix} Y_{11}\Sigma_{12} + (Y_{11}M^* + Y_{12})\Sigma_{22} \\ (MY_{11} + Y_{21})(\Sigma_{12} + M^*\Sigma_{22}) + (MY_{12} + Y_{22})\Sigma_{22} \end{matrix}\Bigg], \tag{3.12.19}$$

where

$$\begin{aligned}
Y_{11} &= X_{11}^{-1} + X_{11}^{-1}X_{12}Y_{22}X_{12}^*X_{11}^{-1}, \\
Y_{12} &= -X_{11}^{-1}X_{12}Y_{22}, \\
Y_{21} &= -Y_{22}X_{12}^*X_{11}^{-1}, \\
Y_{22} &= (X_{22} - X_{12}^*X_{11}^{-1}X_{12})^{-1}, \\
\Sigma_{11} &= (A_{11}^*A_{11} + A_{21}^*A_{21})A_{11}^* + (A_{11}^*A_{12} + A_{21}^*A_{22})A_{12}^*, \\
\Sigma_{12} &= (A_{11}^*A_{11} + A_{21}^*A_{21})A_{21}^* + (A_{11}^*A_{12} + A_{21}^*A_{22})A_{22}^*, \\
\Sigma_{21} &= (A_{12}^*A_{11} + A_{22}^*A_{21})A_{11}^* + (A_{12}^*A_{12} + A_{22}^*A_{22})A_{12}^*, \\
\Sigma_{22} &= (A_{12}^*A_{11} + A_{22}^*A_{21})A_{21}^* + (A_{12}^*A_{12} + A_{22}^*A_{22})A_{22}^*.
\end{aligned}$$

Proof. Consider the block LDL^* decomposition of the full-rank Hermitian matrices $(A^A)^2$, and let us introduce the appropriate unification of some sub-matrices from L and D, if necessary, such that the L and D become 2×2 block matrices. Observing product matrix L^*L^*LDL, the same division, met the following matrix equation:

$$
\begin{aligned}
(L^*LDL^*L)_{11} &= L_{11}^*L_{11}D_{11}L_{11}^*L_{11} + L_{21}^*L_{21}D_{11}L_{11}^*L_{11} + L_{11}^*L_{11}D_{11}L_{21}^*L_{21} \\
&\quad + L_{21}^*L_{21}D_{11}L_{21}^*L_{21} + L_{21}^*L_{22}D_{22}L_{22}^*L_{21} \\
&= D_{11} + L_{21}^*L_{21}D_{11} + D_{11}L_{21}^*L_{21} + L_{21}^*L_{21}D_{11}L_{21}^*L_{21} + L_{21}^*D_{22}L_{21} \\
&= B + B^{-1}CC^*B^{-1}B + BB^{-1}CC^*B^{-1} \\
&\quad + B^{-1}CC^*B^{-1}BB^{-1}CC^*B^{-1} + B^{-1}C(E - C^*B^{-1}C)C^*B^{-1} \\
&= B + B^{-1}CC^* + CC^*B^{-1} + B^{-1}(CC^*B^{-1})^2 + B^{-1}C(E - C^*B^{-1}C)C^*B^{-1} \\
&= B + B^{-1}CC^* + CC^*B^{-1} + B^{-1}CEC^*B^{-1} \\
&= X_{11}, \\
(L^*LDL^*L)_{12} &= L_{11}^*L_{11}D_{11}L_{21}^*L_{22} + L_{21}^*L_{21}D_{11}L_{21}^*L_{22} + L_{21}^*L_{22}D_{22}L_{22}^*L_{22} \\
&= D_{11}L_{21}^* + L_{21}^*L_{21}D_{11}L_{21}^* + L_{21}^*D_{22} \\
&= BB^{-1}C + B^{-1}CC^*B^{-1}BB^{-1}C + B^{-1}C(E - C^*B^{-1}C) \\
&= C + B^{-1}CC^*B^{-1}C + B^{-1}CE - B^{-1}CC^*B^{-1}C \\
&= C + B^{-1}CE = X_{12}, \\
(L^*LDL^*L)_{21} &= (L^*LDL^*L)_{12}^* = X_{12}^*, \\
(L^*LDL^*L)_{22} &= L_{22}^*L_{21}D_{11}L_{21}^*L_{22} + L_{22}^*L_{22}D_{22}L_{22}^*L_{22} \\
&= L_{21}D_{11}L_{21}^* + D_{22} \\
&= C^*B^{-1}BB^{-1}C + E - C^*B^{-1}C \\
&= E = X_{22}.
\end{aligned}
$$

Based on Lemma 3.12.9, invertible matrix X^{-1} is equal to the block matrix

$$
Y = \begin{bmatrix} Y_{11} & Y_{12} \\ Y_{21} & Y_{22} \end{bmatrix}.
$$

Then the following equality are satisfied:

$$\begin{aligned}
LX^{-1}L^* &= \begin{bmatrix} L_{11}Y_{11}L_{11}^* & L_{11}Y_{11}L_{21}^* + L_{11}Y_{12}L_{22}^* \\ L_{21}Y_{11}L_{11}^* + L_{22}Y_{21}L_{21}^* & (L_{21}Y_{21} + L_{22}Y_{21})L_{21}^* + (L_{21}Y_{12} + L_{22}Y_{22})L_{22}^* \end{bmatrix} \\
&= \begin{bmatrix} Y_{11} & Y_{11}L_{21}^* + Y_{12} \\ L_{21}Y_{11} + Y_{21} & (L_{21}Y_{11} + Y_{21})L_{21}^* + L_{21}Y_{12} + Y_{22} \end{bmatrix}, \\
A^*AA^* &= \left[\begin{matrix} (A_{11}^*A_{11} + A_{21}^*A_{21})A_{11}^* + (A_{11}^*A_{12} + A_{21}^*A_{22})A_{12}^* \\ (A_{12}^*A_{11} + A_{22}^*A_{21})A_{11}^* + (A_{12}^*A_{12} + A_{22}^*A_{22})A_{12}^* \end{matrix}\right. \\
&\qquad \left.\begin{matrix} (A_{11}^*A_{11} + A_{21}^*A_{21})A_{21}^* + (A_{11}^*A_{12} + A_{21}^*A_{22})A_{22}^* \\ (A_{12}^*A_{11} + A_{22}^*A_{21})A_{21}^* + (A_{12}^*A_{12} + A_{22}^*A_{22})A_{22}^* \end{matrix}\right] = \begin{bmatrix} \Sigma_{11} & \Sigma_{12} \\ \Sigma_{21} & \Sigma_{22}, \end{bmatrix}.
\end{aligned}$$

Based on the first results of the statement 3.2.7 we have that $A^\dagger = LX^{-1}L^* \cdot A^*AA^*$. Since the equality $L_{21} = M$ is valid, then the following expression is valid (3.12.19) based on product of block matrices $LX^{-1}L^*$ i A^*AA^*. □

For a given block of the matrix $A = \begin{bmatrix} A_{11} & A_{12} \\ A_{21} & A_{22} \end{bmatrix}$, blocks divided matrix $(A^*A)^2 = \begin{bmatrix} B & C \\ C^* & E \end{bmatrix}$ can be easily calculated using the notation,

$$\alpha_{ij} = A_{1i}^*A_{1j} + A_{2i}^*A_{2j},\ i,j = \overline{1,2}.$$

Therefore, blocks B, C, E can be obtained as follows:

$$\begin{aligned}
B &= \alpha_{11}^2 + \alpha_{12}\alpha_{21}, \\
C &= \alpha_{11}\alpha_{12} + \alpha_{12}\alpha_{22}, \\
E &= \alpha_{21}\alpha_{12} + \alpha_{22}^2.
\end{aligned}$$

Calculation of basic matrix $M = C^*B^{-1}$ and $N = CM$ brings us to evaluation block matrices X, using Eq. (3.12.17). Inverse matrix X^{-1} is to be calculated. Note that even the blocks of the matrix Σ can be determined as follows:

$$\Sigma_{11} = \alpha_{11}A_{11}^* + \alpha_{12}A_{12}^*,$$
$$\Sigma_{12} = \alpha_{11}A_{21}^* + \alpha_{12}A_{22}^*,$$
$$\Sigma_{21} = \alpha_{21}A_{11}^* + \alpha_{22}A_{12}^*,$$
$$\Sigma_{22} = \alpha_{21}A_{21}^* + \alpha_{22}A_{22}^*.$$

Finally, Moore–Penrose's inverse $A^\dagger$ can be obtained on the basis of (3.12.19).

Example 3.12.5. Let the next 2×2 block matrix:

$$A = \left[\begin{array}{cc|cc} 5 & 6 & 15 & -6 \\ 16 & 9 & 6 & 9 \\ 7 & 10 & 7 & 13 \\ \hline -5 & 6 & 1 & -6 \\ -6 & -9 & -6 & -19 \\ 7 & 3 & -7 & 12 \end{array}\right].$$

On the basis of Theorem 2.11.8, blocks B C and E matrices $(A^A)^2$ are equal:

$$B = \begin{bmatrix} 505414 & 427211 \\ 427211 & 384895 \end{bmatrix}, C = \begin{bmatrix} 276196 & 664563 \\ 272679 & 549944 \end{bmatrix}, E = \begin{bmatrix} 267870 & 271621 \\ 271621 & 997375 \end{bmatrix},$$

and blocks X_{11}, X_{12} and X_{22} have the following form:

$$X_{11} = \begin{bmatrix} \frac{684308607306455887247574383}{144530479875286494081} & \frac{135121448833328319117306197}{144530479875286494081} \\ \frac{135121448833328319117306197}{144530479875286494081} & \frac{183286313055058544078250686}{144530479875286494081} \end{bmatrix},$$

$$X_{12} = \begin{bmatrix} \frac{6254091839007455}{12022083009} & \frac{2890344581407857}{1335787001} \\ \frac{6969187899173894}{12022083009} & \frac{672434395147319}{1335787001} \end{bmatrix}, X_{22} = \begin{bmatrix} 267870 & 271621 \\ 271621 & 997375 \end{bmatrix}.$$

Inverse matrix awarded X be easily calculated on the basis of Lemma 3.12.9 has the following form:

$$
Y = \left[\begin{array}{cc|cc}
\frac{90644855931319}{1432836657050760392} & -\frac{10229238501285}{1432836657050760392} & \frac{66882803195127683929 7303}{17225681229402306559193379528} & -\frac{24842362463356591191366665}{17225681229402306559193379528} \\
-\frac{10229238501285}{1432836657050760392} & \frac{191635092581707}{716418328525380196} & -\frac{52808391497406171020 18467}{86128406147011532795966 89764} & \frac{81749727214458014224 0559}{17225681229402306559193379528} \\
\hline
\frac{668828031951276839297303}{17225681229402306559193379528} & -\frac{5280839149740617102018467}{8612840614701153279596689764} & \frac{544019002721588448064806883 2588025}{3834973511600883350197925569 7936 52588} & -\frac{37068573518348266128425275392 09653}{230098410696053001011875534187 61915528} \\
-\frac{24842362463356591191366665}{17225681229402306559193379528} & \frac{817497272144580142240559}{17225681229402306559193379528} & -\frac{3706857351834826612842527539209653}{2300984106960530010118755341876191552 8} & \frac{76806172519227726213090282903 23071}{23009841069605300101187553418761915528}
\end{array}\right].
$$

After a few calculations, Moore–Penrose's inverse matrix A is given in next page (1).

Example 3.12.6. Let's now specify rank-deficient matrix A_2 of the previous example and block its LDL^* decomposition of the full-rank. Moore–Penrose's constant inverse matrix L can be easily obtained from:

$$L^{\dagger} = \left[\begin{array}{cc|cc|cc} \frac{11}{21} & \frac{8}{21} & \frac{5}{21} & \frac{2}{21} & -\frac{1}{21} & -\frac{4}{21} \\ \frac{8}{21} & \frac{31}{105} & \frac{22}{105} & \frac{13}{105} & \frac{4}{105} & -\frac{1}{21} \end{array}\right],$$

because of $m = 1$ we have that the following is valid $D^{-1} = D_{11}^{-1} = \begin{bmatrix} -8-x & 9+x \\ 9+x & -10-x \end{bmatrix}$. On the basis of Theorem 2.11.8, generally inverse matrix A_2 can be expressed as given in the next page (2).

Obviously, this approach to find Hermitian matrix is very convenient and efficient; due to the fact that nverse block diagonal matrix D can be easily determined by finding the inverse diagonal matrix D_{ii}, $i = \overline{1,m}$.

Example 3.12.7. To block Hermitian matrix F_6 of Example 2.11.3, the expression obtained in Theorem 3.12.7 can be applied for determination Moore–Penrose's inverse $F_6^{\dagger}$. Block matrix L and D obtained earlier can be taken for evaluation matrix $F_6^{\dagger}$, as

(1)

$$A^{\dagger}=\frac{1}{846414986}\left[\begin{array}{ccc|ccc} 10488670 & 51081296 & -14713647 & -25185490 & 31824819 & 20669699 \\ -5934145 & 14986696 & 18308660 & 93274281 & 7854044 & 25031234 \\ \hline 32955519 & -18409038 & 8264571 & -38970491 & -26726457 & -40470883 \\ -12297891 & -22045404 & 12559500 & -28255629 & -36841766 & -5147046 \end{array}\right].$$

(2)

$$A_2^{\dagger}=\left[\begin{array}{cc|cc|cc} \frac{1}{147}(-8-3x) & \frac{1}{735}(-17-9x) & \frac{2-x}{245} & \frac{1}{735}(29+3x) & \frac{1}{735}(52+9x) & \frac{5+x}{49} \\ \frac{1}{735}(-17-9x) & \frac{-10-9x}{1225} & \frac{25-9x}{3675} & \frac{80+9x}{3675} & \frac{9(5+x)}{1225} & \frac{1}{735}(38+9x) \\ \hline \frac{2-x}{245} & \frac{25-9x}{3675} & \frac{20-3x}{3675} & \frac{5+x}{1225} & \frac{10+9x}{3675} & \frac{1}{735}(1+3x) \\ \frac{1}{735}(29+3x) & \frac{80+9x}{3675} & \frac{5+x}{1225} & \frac{-50-3x}{3675} & \frac{-115-9x}{3675} & \frac{1}{245}(-12-x) \\ \hline \frac{1}{735}(52+9x) & \frac{9(5+x)}{1225} & \frac{10+9x}{3675} & \frac{-115-9x}{3675} & \frac{-80-9x}{1225} & \frac{1}{735}(-73-9x) \\ \frac{5+x}{49} & \frac{1}{735}(38+9x) & \frac{1}{735}(1+3x) & \frac{1}{245}(-12-x) & \frac{1}{735}(-73-9x) & \frac{1}{147}(-22-3x) \end{array}\right].$$

$$
\begin{aligned}
F_6^{\dagger} &= (L^{\dagger})^{*}D^{-1}L^{\dagger} \\
&= \left[\begin{array}{cc|ccc}
1 & 0 & 0 & 0 & 0 \\
0 & 1 & \frac{-4-a}{5+a} & \frac{-3-a}{5+a} & \frac{-2-a}{5+a} \\
\hline
0 & 0 & 1 & 0 & 0 \\
0 & 0 & 0 & \frac{5}{6} & \frac{1}{3} \\
0 & 0 & 0 & \frac{1}{3} & \frac{1}{3} \\
\hline
0 & 0 & 0 & -\frac{1}{6} & \frac{1}{3}
\end{array}\right] \cdot \left[\begin{array}{cc|ccc}
1 & -1 & 0 & 0 & 0 \\
-1 & \frac{6+a}{5+a} & 0 & 0 & 0 \\
\hline
0 & 0 & 2 & -1 & 0 \\
0 & 0 & -1 & -a & 2+a \\
0 & 0 & 0 & 2+a & -3-a
\end{array}\right] \cdot \\
&\quad \left[\begin{array}{cc|ccc|c}
1 & 0 & 0 & 0 & 0 & 0 \\
0 & 1 & 0 & 0 & 0 & 0 \\
\hline
0 & \frac{-4-a}{5+a} & 1 & 0 & 0 & 0 \\
0 & \frac{-3-a}{5+a} & 0 & \frac{5}{6} & \frac{1}{3} & -\frac{1}{6} \\
0 & \frac{-2-a}{5+a} & 0 & \frac{1}{3} & \frac{1}{3} & \frac{1}{3}
\end{array}\right] \\
&= \left[\begin{array}{cc|ccc|c}
1 & -1 & 0 & 0 & 0 & 0 \\
-1 & 2 & -1 & 0 & 0 & 0 \\
\hline
0 & -1 & 2 & -\frac{5}{6} & -\frac{1}{3} & \frac{1}{6} \\
0 & 0 & -\frac{5}{6} & \frac{7}{9} - \frac{a}{4} & \frac{4}{9} & \frac{1}{9} + \frac{a}{4} \\
0 & 0 & -\frac{1}{3} & \frac{4}{9} & \frac{1}{9} & -\frac{2}{9} \\
\hline
0 & 0 & \frac{1}{6} & \frac{1}{9} + \frac{a}{4} & -\frac{2}{9} & -\frac{5}{9} - \frac{a}{4}
\end{array}\right].
\end{aligned}
$$

3.13 Greville's Partitioning Method

Currently, Leverrier–Faddeev algorithm [29] and partitioning methods [21]) are usually taken for calculation of Moore–Penrose's inverse. Various modifications of "partitioning" algorithm papers, polynomial and rational matrices are made and tested in the Ref. [65].

The first partitioning method was introduced in [21]. This method is a recursive algorithm for calculating the Moore–Penrose's inverse. In this method the expression that connects the Moore–Penrose's inverse matrix $[A|a]$ (A matrix with added column a) and Moore–Penrose's inverse matrix A are used. The initial case is Moore–Penrose's inverse. First column a_1

matrix A, which is calculated according to the formula

$$A_1^\dagger = a_1^\dagger = \begin{cases} \frac{1}{a_1^* a_1} a_1^*, & a_1 \neq 0 \\ a_1^* & a_1 = 0. \end{cases} \tag{3.13.1}$$

Further, for each column $a_i, 2 \leq i \leq n$ of matrix $A \in \mathbb{C}_r^{m \times n}$ we have the following. If,

$$A_i = [A_{i-1} | a_i],$$

then,

$$A_i^\dagger = \begin{bmatrix} A_{i-1}^\dagger - d_i b_i^* \\ b_i^* \end{bmatrix}, \quad 2 \leq i \leq n$$

where,

$$d_i = A_{i-1}^\dagger a_i,$$

$$c_i = a_i - A_{i-1} d_i = (I - A_{i-1} A_{i-1}^\dagger) a_i,$$

$$b_i = \begin{cases} \frac{1}{c_i c_i^*} c_i & c_i \neq 0 \\ \frac{1}{1 + d_i^* d_i} (A_{i-1}^\dagger)^* d_i & c_i = 0. \end{cases} \tag{3.13.2}$$

Finally, $A^\dagger = a_n^\dagger$. In Ref. [65], Stanimirović and Tasić described an algorithm for accurately calculating the Moore–Penrose's inverse, which is based on the applied residual arithmetic and such a method alterations.

Example 3.13.1. Let be given the matrix

$$A = \begin{bmatrix} 1 & 3 \\ 2 & 2 \\ 3 & 1 \end{bmatrix}.$$

$$A_1^\dagger = a_1^\dagger = \frac{1}{a_1^* a_1} a_1^* = \frac{1}{14} \begin{bmatrix} 1 & 2 & 3 \end{bmatrix}.$$

$$d_2 = A_1^\dagger a_2 = \frac{1}{14} \begin{bmatrix} 1 & 2 & 3 \end{bmatrix} \begin{bmatrix} 3 \\ 2 \\ 1 \end{bmatrix} = \frac{5}{7}.$$

$$c_2 = a_2 - A_1 d_2 = \begin{bmatrix} 3 \\ 2 \\ 1 \end{bmatrix} - \begin{bmatrix} 1 \\ 2 \\ 3 \end{bmatrix} \cdot \frac{5}{7} = \frac{4}{7} \cdot \begin{bmatrix} 4 \\ 1 \\ -2 \end{bmatrix}$$

$$b_2 = \frac{1}{c_2^* c_2} c_2 = \frac{1}{12} \begin{bmatrix} 4 \\ 1 \\ -2 \end{bmatrix}$$

$$A^\dagger = A_2^\dagger = \begin{bmatrix} A_1^\dagger - d_2 b_2^* \\ b_2^* \end{bmatrix} = \begin{bmatrix} -\frac{1}{6} & \frac{1}{12} & \frac{1}{3} \\ \frac{1}{3} & \frac{1}{12} & -\frac{1}{6} \end{bmatrix}$$

Greville suggested recursive algorithm that connects pseudoinverse R matrix supported by the appropriate vector r and pseudoinverse $R^\dagger$ of R. Let A be a matrix of dimension $m \times n$. Let A_i be sub-matrix of A containing i the first column of the matrix A. If i-thcolumn of matrix A is denoted by a_i, then A_i granted as $A_i = [A_{i-1} \mid a_i]$.

Algorithm 3.12 Greville's partitioning method

1: Starting values:

$$A_1^\dagger = a_1^\dagger = \begin{cases} \frac{1}{a_1^* a_1} a_1^*, & a_1 \neq 0 \\ a_1^* & a_1 = 0. \end{cases} \tag{3.13.3}$$

2: The recursive step: For each column $a_i, 2 \leq i \leq n$ from A we calculate

$$A_i^\dagger = \begin{bmatrix} A_{i-1}^\dagger - d_i b_i^* \\ b_i^* \end{bmatrix}, \quad 2 \leq i \leq n$$

where

2.1: $d_i = A_{i-1}^\dagger a_i$,

2.2: $c_i = a_i - A_{i-1} d_i = (I - A_{i-1} A_{i-1}^\dagger) a_i$,

2.3:

$$b_i = \begin{cases} \frac{1}{c_i c_i^*} c_i & c_i \neq 0 \\ \frac{1}{1 + d_i^* d_i} (A_{i-1}^\dagger)^* d_i & c_i = 0. \end{cases} \tag{3.13.4}$$

3: Stopping condition: $A^{\dagger} = A_n^{\dagger}$.

Example 3.13.2. For the matrix $A = \begin{bmatrix} -1 & 0 & 1 & 2 \\ -1 & 1 & 0 & -1 \\ 0 & -1 & 1 & 3 \\ 0 & 1 & -1 & -3 \\ 1 & -1 & 0 & 1 \\ 1 & 0 & -1 & -2 \end{bmatrix}$ we get

$$A^{\dagger} = \begin{bmatrix} -\frac{5}{34} & -\frac{3}{17} & \frac{1}{34} & -\frac{1}{34} & \frac{3}{17} & \frac{5}{34} \\ \frac{4}{51} & \frac{13}{102} & -\frac{5}{102} & \frac{5}{102} & -\frac{13}{102} & -\frac{4}{51} \\ \frac{7}{102} & \frac{5}{102} & \frac{1}{51} & -\frac{1}{51} & -\frac{5}{102} & -\frac{7}{102} \\ \frac{1}{17} & -\frac{1}{34} & \frac{3}{34} & -\frac{3}{34} & \frac{1}{34} & -\frac{1}{17} \end{bmatrix}.$$

Example 3.13.3. For the matrix $A = \begin{bmatrix} -1 & 0 & 1 & 2 \\ -1 & 1 & 0 & -1 \\ 0 & -1 & 1 & 3 \\ 1 & 1 & -2 & -5 \end{bmatrix}$ we get

$$A^{\dagger} = \begin{bmatrix} -\frac{11}{51} & -\frac{6}{17} & \frac{7}{51} & \frac{4}{51} \\ \frac{7}{51} & \frac{13}{51} & -\frac{2}{17} & -\frac{1}{51} \\ \frac{4}{51} & \frac{5}{51} & -\frac{1}{51} & -\frac{1}{17} \\ \frac{1}{51} & -\frac{1}{17} & \frac{4}{51} & -\frac{5}{51} \end{bmatrix}.$$

3.13.1 Another Method of Alterations

A similar algorithm of alterations comes from Matveeva. In Matveeva paper, an algorithm is given based on the sequential calculations of Moore–Penrose's inverse of the initial matrix calculated using Moore–Penrose's inverse. The base case is the Moore–Penrose inverse of the first kind.

$$A_1 = a_1^T, \quad A_{k+1} = \begin{bmatrix} A_k \\ a_{k+1}^T \end{bmatrix}.$$

The formula for U_{k+1} represents the Moore–Penrose inverse of sub-matrix which contains the first $k+1$ alterations of the starting matrix, a_{k+1}^T is $k+1$-th row of the matrix, and v_{k+1} is $k+1$-th column.

The algorithm is based on the following recurrent relations:

$$
\begin{aligned}
&(1) \qquad U_{k+1} = \left[U_k - v_{k+1} - v_{k+1}a_{k+1}^T | v_{k+1}\right],\\
&(2) \qquad U_1 = (a_1^T a_1)^{-1} a_1,\\
&(3) \qquad v_{k+1} = T_k a_{k+1}(a_{k+1}^T T_k a_{k+1} + 1)^{-1},\\
&(4) \qquad v_{k+1} = S_k a_{k+1}(a_{k+1}^T S_k a_{k+1} + 1)^{-1},\\
&(5) \qquad T_{k+1} = T_k - v_{k+1} a_{k+1}^T T_k,\\
&(6) \qquad T_1 = I - U_1 A_1 = I - U_1 a_1^T,\\
&(7) \qquad S_{k+1} = S_k - v_{k+1} a_{k+1}^T S_k,\\
&(8) \qquad S_k = U_k U_k^T = (A_k^T A_k)^{-1}.
\end{aligned}
$$

Note that the formula (3) is used if A_{k+1}^T matrix increases rank a_k, the formula (4) is used when A_{k+1}^T does not increase the rank of the matrix a_k. The checks are not simple. Therefore, the application of this method is not recommended.

The Zhuhovski paper is used for the following codes as the matrix $A\ in\ mathbbC_r^{m\times n}$.

$a_t, \quad t = 1,\ldots,m,$ – kind of matrix A;

$A_t = \begin{bmatrix} a_1 \\ \ldots \\ a_t \end{bmatrix}, \quad t = 1,\ldots,m$ – sub-matrix dimensions $t \times n$, formed from the first type t matrix A; $y^t = A_t x$, and especially

$y_t = a_t x, \quad t = 1,\ldots,m;$

$x_t = A_t^\dagger y^t, \quad t = 1,\ldots,m;$

$\gamma_t = I_n - A_t^\dagger A_t, \quad t = 1,\ldots,m.$

3.13.2 Zhuhovski Method

Starting from the known recurrent equations for solving the linear system $Ax = y$, where $A \in \mathbb{C}_m^{m\times n}$.

$$x_{t+1} = x_t + \frac{\gamma_t a_{t+1}^*}{a_{t+1}\gamma_t a_{t+1}^*}\left(y_{t+1} - a_{t+1}x_t\right), \quad x_0 = \vec{0},$$

$$\gamma_{t+1} = \gamma_t - \frac{\gamma_t a_{t+1}^* a_{t+1}\gamma_t}{a_{t+1}\gamma_t a_{t+1}^*}, \quad \gamma_0 = I_n.$$

Zhuhovski introduced the generalization of these recurrent relations, and found a solution for consensual system of linear algebraic equation $Ax = y$, in the case of the matrix $A \in \mathbb{C}_r^{m\times n}$, $r \le m \le n.$, it was shown that X_T, $\gamma_t, t = 1,\ldots,m$ are solutions gained following the system of equations:

$$x_{t+1} = x_t + \gamma_t a_{t+1}^* \left(a_{t+1}\gamma_t a_{t+1}^*\right)^\dagger \left(y_{t+1} - a_{t+1}x_t\right), \quad x_0 = \vec{0},$$

$$\gamma_{t+1} = \gamma_t - \gamma_t a_{t+1}^* \left(a_{t+1}\gamma_t a_{t+1}^*\right)^\dagger a_{t+1}\gamma_t, \quad \gamma_0 = I_n,$$

where

$$\left(a_{t+1}\gamma_t a_{t+1}^*\right)^\dagger = \begin{cases} \frac{1}{a_{t+1}\gamma_t a_{t+1}^*} & a_{t+1}\gamma_t a_{t+1}^* > 0, \\ 0 & a_{t+1}\gamma_t a_{t+1}^* = 0. \end{cases} \tag{3.13.5}$$

Note that the equation $a_{t+1}\gamma_t a_{t+1}^* = 0$ if and only if row a_{t+1} of linearly dependent on the type of $a_1,\ldots,a_t$.

Let $\Gamma_t, \quad t = 1,\ldots,n$ for $n\times n$ matrix, defined using:

$$\Gamma_{t+1} = \Gamma_t - \Gamma_t b_{t+1}\left(b_{t+1}^*\Gamma_t b_{t+1}\right)^\dagger b_{t+1}^*\gamma_t, \quad \gamma_0 = I_n.$$

Now, let $b_t, \quad t = 1,\ldots,n$denote columns of matrix A, while $c_t, \quad t = 1,\ldots,n$ are rows of matrix $I_n - \Gamma_n$. Consider the array of matrices X_t and γ_t dimension $n\times n$, defined as follows:

$$\gamma_{t+1} = \gamma_t - \gamma_t a_{t+1}^* \left(a_{t+1}\gamma_t a_{t+1}^*\right)^\dagger a_{t+1}\gamma_t, \quad \gamma_0 = I_n,$$

$$X_{t+1} = X_t + \gamma_t a_{t+1}^* \left(a_{t+1}\gamma_t a_{t+1}^*\right)^\dagger (c_{t+1} - a_{t+1}x_t), \quad X_0 = \mathbb{O}.$$

3.14 Leverrier–Faddeev Method for Computing Moore–Penrose Inverse of Polynomial Matrices

The Leverrier–Faddeev method (Algorithm 2.2), presented and described in Subsection 2.2.2 is applicable to rational and polynomial matrices $A(s)$. However, final expression for $A^\dagger(s)$ will be undefined for the zeros of polynomial (rational function) $a_k(s)$. A generalized inverse for these values of argument s can be computed independently by Algorithm 2.2 or by some other method.

We will investigate a complexity analysis of the Algorithm 2.2 when an input matrix is polynomial. For the sake of simplicity denote $A'(s) = A(s)A^T(s)$ and $d' = \deg A'(s)$. Since $A(s) \in \mathscr{R}[s]^{n\times m}$, there holds $A'(s) \in \mathscr{R}[s]^{n\times n}$. Therefore, in step 5 we need to multiply two matrices of the order $n \times n$. This multiplication can be done in time $\mathscr{O}(n^3)$ when A is a constant matrix, but in the polynomial case corresponding time is $\mathscr{O}\left(n^3 \cdot d' \cdot \deg B_{j-1}(s)\right)$.

Definition 3.14.1. *Define $k^{A(s)}$, $a_i^{A(s)}$ and $B_i^{A(s)}$, as the values of k, $a_i(s)$ and $B_i(s)$, $i = 0,\dots,n$, computed by Algorithm 2.2 when the input matrix is constant, rational or polynomial matrix $A(s)$. Also denote $a^{A(s)} = a_{k^{A(s)}}^{A(s)}$ and $B^{A(s)} = B_{k^{A(s)}-1}^{A(s)}$.*

It can be proved by the mathematical induction that holds the inequality $\deg B_j(s) \le j \cdot d'$ for $j = 0,\dots,n$ where the equality is reachable. Then the required time for the matrix–matrix multiplication in step 5 is $\mathcal{O}(n^3 \cdot j \cdot d'^2)$. Similarly, one can verify that the required time for steps 6 and 7 is $\mathcal{O}(n \cdot j \cdot d')$. Therefore, the total running time of loop body in steps 4–8 is $\mathcal{O}(n^3 \cdot j \cdot d'^2)$.

$$\mathcal{O}\left(\sum_{j=1}^{n} n^3 \cdot j \cdot d'^2\right) = \mathcal{O}(n^5 \cdot d'^2) \tag{3.14.6}$$

Here we can see that complexity of Algorithm 2.2 for polynomial matrices is n times bigger than in the case of constant matrices (subsection 2.2.5). Practically, this complexity is smaller than (3.14.6) (not all elements of matrices $B_j(s), A_j(s)$ and $A'(s)$ has the maximal degree), but it is still large.

Note that for sparse input matrices $A(s)$ there are known modifications of Algorithm 2.2. Most of them are done by Karampetakis [30, 32], Stanimirovic [62] and also by Stanimirovic and Karampetakis [63]. Instead of these methods, interpolation method presented in this section is designed both for dense and sparse matrices.

CHAPTER 4

APPLICATIONS

Generalized inverses are very powerful tools and are applied in many branches of mathematics (also in other sciences and technics). We have already seen some applications in solution of the matrix equations (Section 2.1). There are a lot of other disciplines in which generalized inverses plays the important role. For example, estimation theory (regression), computing polar decomposition, the areas of electrical networks, automatic control theory, filtering, difference equations, pattern recognition, etc. It is worth mentioning that the main application of generalized inverses is the linear regression model, discussed in this section. Note that more detailed consideration of this state-of-art application is given in the monograph [3]. In this chapter, we will illustrate two applications, in mathematical statistics (linear regression) and automatic control (feedback compensation problem). We hope that these two applications illustrate the richness and potential of the generalized inverses.

4.1 The Problem of Feedback Compensation

In this subsection we will show one application of the generalized inverses theory in automatic control. First, we will make very brief introduction in automatic control and linear system theory. An automatic control system receives the input signal $x(t)$ and produces the output signal $y(t)$.

If both input and output signal are supposed to be continuous functions, then we say that corresponding control system is *continuous*. The continuous system is called *causal* if $x(t) = 0$ for all t implies $y(t) = 0$ for all t. In other words, zero input of the causal system produces the zero output. In the rest of the subsection we will suppose causality of all considered dynamic systems.

For the continuous system it is usual to consider input and output signal in the Laplace domain (or s-domain). In other words, instead of working with functions $x(t)$ and $y(t)$ we will consider their Laplace transforms

$$X(s) = \int_0^{+\infty} x(t)e^{-st}dt, \quad Y(s) = \int_0^{+\infty} y(t)e^{-st}dt. \tag{4.1.1}$$

Every system can have several inputs and several outputs and therefore $X(s)$ and $Y(s)$ (also $x(t)$ and $y(t)$) are vectors in general case. We will suppose that $X(s)$ and $Y(s)$ are vectors of the format $m \times 1$ and $n \times 1$. For the linear dynamic system, signals $X(s)$ and $Y(s)$ are connected by linear relations, i.e., then there exists matrix $G(s)$, called *transfer matrix* such that holds

$$Y(s) = G(s)X(s).$$

Hence, every linear dynamic system is fully described by its transfer matrix $G(s)$. Additional assumption on $G(s)$ is that it is rational or polynomial matrix. In such case, relation $Y(s) = G(s)X(s)$ in time domain represents the system of integro-differential relations.

Linear dynamic systems are usually represented by block diagrams. Several blocks can be combined in the construction of more complex linear systems.

Consider the open-loop system where $G(s)$ is $m \times n$ rational or polynomial transfer matrix. The problem is to determine if there exists feedback block, described by rational transfer matrix $F(s) \in \mathscr{C}^{n\times m}(s)$ such that closed loop system has desired transfer matrix $H(s) \in \mathscr{C}^{m\times n}(s)$.

It holds $H(s) = (I_m + G(s)F(s))^{-1}G(s)$ and therefore the following equation must be satisfied

$$G(s)F(s)H(s) = G(s) - H(s), \tag{4.1.2}$$

where $F(s)$ is unknown matrix. The next theorem gives us the sufficient and necessary condition for existence of the solution of feedback compensation problem for given matrices $G(s)$ and $H(s)$.

Theorem 4.1.1. *The sufficient and necessary condition for solvability of the matrix equation (4.1.2) is given by*

$$G(s) = G(s)H(s)^{\dagger}H(s), \quad H(s) = G(s)G(s)^{\dagger}H(s). \tag{4.1.3}$$

Proof. Applying the Theorem 2.1.17 we obtain the following necessary and sufficient condition for existence of the solution of equation (4.1.2)

$$G(s)G(s)^{\dagger}(G(s) - H(s))H(s)^{\dagger}H(s) = G(s) - H(s). \tag{4.1.4}$$

Last relation can be further simplified and expressed in the following form

$$G(s)(I_n - H(s)^{\dagger}H(s)) = (I_m - G(s)G(s)^{\dagger})H(s). \tag{4.1.5}$$

By multiplying (4.1.5) with $H(s)^{\dagger}$ on the right side we obtain

$$(I_m - G(s)G(s)^{\dagger})H(s)H(s)^{\dagger} = G(s)(H(s)^{\dagger} - H(s)^{\dagger}H(s)H(s)^{\dagger}) = \mathbb{O}$$

and therefore holds

$$H(s)H(s)^{\dagger} = G(s)G(s)^{\dagger}H(s)H(s)^{\dagger}. \tag{4.1.6}$$

Again multiplying (4.1.6) with $H(s)$ from right side yields

$$H(s) = G(s)G(s)^{\dagger}H(s). \tag{4.1.7}$$

It can be analogously proved that

$$G(s) = G(s)H(s)^{\dagger}H(s). \tag{4.1.8}$$

Note that relations (4.1.7) and (4.1.8) imply relation (4.1.5) in such way that both sides of (4.1.7) are equal to $\mathbb{O}$. This proves that (4.1.5) is equivalent with the system (4.1.7) and (4.1.8). □

In order to check the condition (4.1.3) we have to compute MP inverses $G(s)^{\dagger}$ and $H(s)^{\dagger}$. This computation can be done by any of the algorithms from the previous sections. Note that if $A(s) \in \mathscr{C}^{m\times n}(s)$ is rational matrix, it can be represented as

$$A(s) = \frac{1}{a_1(s)}A_1(s) \tag{4.1.9}$$

where $A_1(s) \in \mathscr{C}^{m\times n}[s]$ is polynomial matrix and polynomial $a_1(s)$ is equal to the least common multiplier of all denominators in $A(s)$. Now the MP inverse $A(s)^{\dagger}$ can be represented as $A(s)^{\dagger} = a_1(s)A_1(s)^{\dagger}$. Using the relation (4.1.9) we can reduce the computation of generalized inverse of rational matrix $A(s)$ to the same computation for polynomial matrix $A_1(s)$. The same holds also for Drazin and some other classes of generalized inverses. This fact enables us to use methods developed for the polynomial matrices in computation of rational matrix generalized inverses.

If (4.1.3) is satisfied, that is, there exists the solution of feedback problem, according to the Theorem 2.1.17 all solutions of (4.1.2) are given by

$$F(s) = G(s)^{\dagger}(G(s)-H(s))H(s)^{\dagger} + Y(s) - G(s)^{\dagger}G(s)Y(s)H(s)H(s)^{\dagger}. \tag{4.1.10}$$

Among the all matrices $F(s)$ defined by (4.1.10) we have to select only those such that matrix $I_m + G(s)F(s)$ is regular (otherwise, closed-loop feedback system oscillates and output $y(t)$ is independent from input $x(t)$).

Example 4.1.1. *[30]*

Consider the following transfer matrix of the open loop system

$$G(s) = \begin{bmatrix} \frac{1}{s-2} & 0 & 0 \\ 0 & \frac{1}{s-1} & 0 \end{bmatrix}.$$

We would like to find out if there exists the feedback such that closed loop system has the following transfer matrix

$$H(s) = \begin{bmatrix} 0 & \frac{1}{s+1} & 0 \\ \frac{1}{(s+1)^2} & 0 & 0 \end{bmatrix}.$$

First we compute generalized inverses $G(s)^{\dagger}$ and $H(s)^{\dagger}$. These values are

$$G(s) = \begin{bmatrix} s-2 & 0 \\ 0 & s-1 \\ 0 & 0 \end{bmatrix}, \qquad H(s) = \begin{bmatrix} 0 & (s+1)^2 \\ s+1 & 0 \\ 0 & 0 \end{bmatrix}.$$

Direct verification yields that relations $H(s) = G(s)G(s)^{\dagger}H(s)$ and $G(s) = G(s)H(s)^{\dagger}H(s)$ are satisfied. According to the Theorem 4.1.1, there exists the solution of the feedback problem. We will use expression (4.1.10) to determine value of $F(s)$. Set $Y(s) = \mathbb{O}$. Obtained solution is

$$F(s) = G(s)^{\dagger}(G(s) - H(s))H(s)^{\dagger} = \begin{bmatrix} 2-s & (s+1)^2 \\ s+1 & 1-s \\ 0 & 0 \end{bmatrix}.$$

Also by direct verification we conclude that matrix $I_2 + G(s)F(s)$ is regular and therefore $F(s)$ is valid solution of the feedback problem.

At the end of this section let us mention that similar approach can be applied on the more complex automatic control systems.

4.2 Linear Regression Method

In many sciences, very often there is a need determination direct functional dependence between two or more sizes. Oblast mathematical statistics dealing with the determination and describing this dependence is called regression. If the determination linear relationship observed insignia, it is a linear regression. There are two possible approaches to this problem.

In the first approach Looking at the impact insignia (random size) $X_1,\dots,X_n$ the also random size Y.At the same time, to determine the function $f(x_1,\dots,x_n)$ such that random size $f(X_1,\dots,X_n)$ best approximates Y. As a measure of deviation of these two sizes, usually mean square deviation is used, i.e., provided that the $E(Y-f(X))$ minimum. This type of regression will not be addressed. In the second approach, looks at the impact of a certain number of non-random size, control factor to a value appropriate random size Y. This connection is deterministic (non-random). The size of the Y also affect and random factors, as well as other non-random factors the impact can not be effectively seen. We'll assume that the impact of these two mutually independent, additive (value of the variable Y is the sum of deterministic and random components) and expectation that the random component is equal to zero.

Let $a_1,\dots,a_p$ the value of the control factor. We will assume that these factors affect the Y function through $f_i(a_1,\dots,a_p)$ where $i=1,\dots,n$, whereby the dependence of the size of Y of functions $f_i value$ linear. In other words, we will assume that the vase

$$Y = x_1 f_1(a_1,\dots,a_p) + \dots + x_n f_n(a_1,\dots,a_p) + \varepsilon. \tag{4.2.11}$$

With $x_1,\dots,x_n$, we highlight the value of the coefficient of linear bonds, and with ε random component. This model is called linear regression model other species. It is necessary to carry out an appropriate assessment of the odds $x_1,\ldots,x_n$ so that relation (4.2.11) "best" agrees with the obtained experimental results. When it is carried out a series of experiments with different values of the control factors $a^k = (a_{k1},\ldots,a_{kn})$ $k = 1,\ldots,m$ where m is the total number of experiments. Denote by Y_i and ε_i, $i = 1,\ldots,m$ the resulting value of Y size and random size ε in i -th experiment (repetition). And is obtained by

$$Y_i = x_1 f_1(a_{i1},\ldots,a_{ip}) + \ldots, x_n f_n(a_{i1},\ldots,a_{ip}) + \varepsilon_i, \quad i = 1,\ldots,m. \tag{4.2.12}$$

Denote now $\mathbf{Y} = [Y_1 \ \cdots \ Y_m]^T$, $A = [f_j(a_{i1},\ldots,a_{ip})]$, $x = [x_1 \ \cdots \ x_n]^T$ and denote $\varepsilon = [\varepsilon_1 \ \cdots \ \varepsilon_m]^T$. The system of equations (4.2.12) can be written in matrix form as follows

$$\mathbf{Y} = \mathbf{Ax} + \varepsilon \tag{4.2.13}$$

It is necessary to specify the value of the matrix A and vector Y find the value of the vector $\mathbf{x} = \mathbf{x}^*$ such that the norm of the vector ε is minimal, i.e.,

$$\min_{\mathbf{x}\in\mathscr{R}^{n\times 1}} \|\mathbf{Y} - \mathbf{Ax}\| = \|\mathbf{Y} - \mathbf{Ax}^*\|. \tag{4.2.14}$$

Depending on the selection norms $\|\cdot\|$ differentiate between different types of second order linear regression. Most often taken Euclidean L_2 norm, but are used and L_1, L_p and L_∞ norms. As for the L_1 and L_∞ norm, problem (4.2.14) boils down to the problem of linear programming.

Assume that the norm in (4.2.14) Euclidean. Problem (4.2.14) is the problem of finding the minimal medium square solution system of equations $A\mathbf{x} = \mathbf{Y}$. Based on the Theorem 2.1.9, we have a solution that $\mathbf{x}^*$ determined following expression

$$\mathbf{x}^* = A^\dagger \mathbf{Y} + (I_m - A^\dagger A)z, \quad z \in \mathscr{R}^{m\times 1} \tag{4.2.15}$$

Therefore, to determine x^*, it is enough to calculate $A^\dagger$. Note that if the matrix A the full-rank of species, and if rank$A = m$, then on the basis of Lemma 2.1.12 is valid $A^\dagger = (A^*A)^{-1}A^*$, and $A^\dagger A = I_m$. However, the term (4.2.15) down to

$$\mathbf{x}^* = (A^*A)^{-1}A^*\mathbf{Y}. \tag{4.2.16}$$

Example 4.2.1. *Assume that the size of Y depends on a control factor t and is a form of addiction*

$$Y = at + bt^3 + ct^5 + \varepsilon. \tag{4.2.17}$$

Here we have been given the following set of values:

t_i	-1	-0.5	0.5	1
Y_i	-8.98112	-1.00373	1.04187	9.01672

First, we form the matrix A and vector $\mathbf{Y}$

$$A = \begin{bmatrix} t_1 & t_1^3 & t_1^5 \\ t_2 & t_2^3 & t_2^5 \\ t_3 & t_3^3 & t_3^5 \\ t_4 & t_4^3 & t_4^5 \end{bmatrix} = \begin{bmatrix} -1 & -1 & -1 \\ -0.5 & -0.125 & -0.03125 \\ 0.5 & 0.125 & 0.03125 \\ 1 & 1 & 1 \end{bmatrix}, \quad \mathbf{Y} = \begin{bmatrix} -8.98112 \\ -1.00373 \\ 1.04187 \\ 9.01672 \end{bmatrix}$$

Then we compute the MP inverse $A^\dagger$, and solution application (4.2.15) where the $z = \mathbb{O}$, smallmatrix

$$A^\dagger = \begin{bmatrix} 0.0833333 & -1.14286 & 1.14286 & -0.0833333 \\ -0.25 & 0.380952 & -0.380952 & 0.25 \\ -0.333333 & 0.761905 & -0.761905 & 0.333333 \end{bmatrix}, \quad \mathbf{x}^* = \beta$$

where $\beta = A^\dagger\mathbf{Y} = \begin{bmatrix} 0.838005 \\ 3.72018 \\ 4.44073 \end{bmatrix}$.

Therefore, the best L_2 approximation of the size of Y form (4.2.17) is

$$Y = 0.838005t + 3.72018t^3 + 4.44073t^5$$

4.3 Restoration of Blurred Images

In this section, we will deal with the renewal of damaged digital images provided by use camera or camcorder. When using the camera, we want to be faithful to the recorded image see the scenes that we see; however, we have that every image is more or less blurred. Thus, restoration process is of fundamental importance to obtain sharpness and clarity of images. A digital image is made up of elements called pixels. Each pixel is assigned intensity, which should be characterized by the color of the small rectangular segment of the scene you paint. Photographs of small size usually have about $256^2 = 65,536$ pixels, while on the other side of the picture to high resolution have 5 to 10 million pixels. When shooting a scene for obtaining digital image must contact a blur, because it is impossible to avoid the so-called “spillover” information from the original scene on adjacent pixels. For example, the optical system of the camera lens may be out of focus, so that we have that the incoming light blurred. The same problem occurs, for example, when processing images in astronomy, where the incoming light into the telescope goods distorted because of turbulence present in the air. In these and similar situations, the inevitable result is obtaining a blurred image.

For the restoration of images, we want to restore the original, clear, image using a mathematical model that we present blurring process. The key question is whether some lost information contained in degraded picture If so, this information is in some meaning “hidden” and can only be recovered if we know the details of the blurring process. Unfortunately, there is no hope that we can rebuild the original image to the smallest detail! That due to various errors that are inevitable in the process of recording and imaging. The most important thing mistakes were very instability of the image capturing process, and errors in approximations, or presentation of images by means of a finite number of digits.

One of the challenges of the restoration of paintings is to design efficient and reliable algorithm to obtain as much information as possible about the original image, on the basis of the data provided. In this chapter, we will consider just this problem with images in which there has been a blur due to the uniform linear motion camera or scene you want to picture taken. In the sciences such as physics and mathematics, is usually taken to be valid assumption of linear blur because in many real situations blur really linear, or at least some can be approximated by a linear model. This assumption is very useful because it is known to have a number of tools in linear algebra and from matrix accounts. Using linear algebra in the reconstruction process image has a long history, and reaches back to classic works such as books of Andrews and Hunt.

Pictures provide very useful information, however, is not a rare case of existence blur (Blurry contours). Due to the relative motion between the camera and the object (target) that should be photographed blurring caused by motion (motion blur). Restoration (reconstruction) image blurred some movement for many years is one of the main problems in the field of digital image processing. Recover damaged images for as much copies of the initial the original is of great significance and is used in many fields such as medicine, military control, satellite and astronomical image processing, remote sensing, etc. As expected, very much scientific interest is in the area of image restoration (see [11, 27, 30, 31, 73]). It is known that every digital black-and-white image we present two dimensional using series (rectangular matrix), in the sense that each pixel corresponds to a number that is actually element of this matrix. In this way, the problem of the restoration of the image is shifted to the field of mathematics, or precisely the manipulation of matrices and linear algebra. In this doctorate will deal eliminating the blur with the image that was created due to the uniform linear motion. It is assumed that this linear movement closely linked to the whole number of pixels and be seen horizontal and vertical sampling. Moreover, we have to be degraded (blurred) picture (which usually denoted by G) can be represented as a convolution of the original image (which will mark with F) and function expansion point (point-spread function PSF), which is also known as the core of blur (denote it as H). For uniform linear motion to the core blur, can be represented by means of a characteristic Toeplitz's matrices.

And we'll call an unknown parameter blur. Blur caused by horizontal or vertical linear motion is modeled using the matrix equation HT F=G and HF, respectively, compared to the unknown (requested) matrix F. Clearly, Moore–Penrose inverse matrix $H\dagger$, which is the original cause of blur matrix F is used to find a solution to these equations [30, 31]. Based on the well-known fact that the inverse Moore–Penrose useful tool for solving a system of linear matrix equations [12, 112] we have a large influence application of the inverse of the power restoration image [20, 3032]. The use of matrix pseudo inverse in the process of image reconstruction is one of the most common approaches in solving this problem [20]. In other words, the basic problem with which we have to face is the selection of an efficient algorithm for calculating the generalized inverse H. Algorithms that are used in the Refs. [30, 31] for H calculation is based on a method for finding Moore–Penrose's complete inverse matrix rank, have been introduced in Refs. [87, 88]. Approximation of the original image obtained in the Ref. [30] are reliable and very accurate.

In the literature, there are different methods for calculating direct-Moore–Penrose's inverse (see, for example, [12, 129]). P. Courrieu in their work [36] proposes an algorithm for fast calculation Moore–Penrose's inverse, which is based on the principle of reverse sequence (reverse order law) and Cholesky factorization on the full-rank singular symmetric matrices. Also, an efficient method for calculating the Moore–Penrose's inverse matrix rectangular complete ranking and a square matrix with at least one zero column or type is presented in Refs. [87, 88]. Greville in his paper [61] gives a recurrent rule for calculating the Moore–Penrose's inverse. Later, Udwadia and Kalaba provide a simpler and different evidence Greville's formula [141]. Because features sequential calculation Greville's partitioning method is widely used in filtering theory, the theory of linear evaluation, system identification, optimization and field analytical dynamics [59, 76, 81, 82, 120]. In contrast to such method where the added recursively column after column (species by species) Bhimsankaram is a method in which the step added blocks matrix [13]. The author in his work as evidence freely verify that the output matrix meets four equations (conditions) for the Moore–Penrose inverse.

Plain partitioning method gives the best results for small the value of the parameter l. It is obvious that the CPU time partitioning method is linearly growing in relation to the parameter of blur so it is expected that at some point (for sufficiently large l) becomes greater than all other observed methods. Therefore, we conclude that the plain partitioning method is acceptable only for small values of the parameter degradation l = 20. For values l ¿ 20 it is clear that the block partitioning method gives the best results so far in relation to the ordinary partitioning and in relation to other methods observed. As we know the block partitioning method requires only one recursive step to finding H, while Greville's method requires l-1 steps. The overall conclusion is that the practical complexity of calculation, obtained on the basis of implementation incorporated in the

MATLAB package, conversely proportional to the size of the block matrix. As a consequence of all the above, we choose the block partitioning methods as efficient and the observers analysis below and compare it with other suggested methods.

Blur generated when forming the X-ray image(s) usually occur as the result of moving the camera or the object that needs to take a photo. Problem restoration these images can be very effectively modeled using blur obtained previously described uniform linear motion. Figure 3.2 represents a practical example of restoration blurred X-ray images. The original picture was taken from the Internet after a search on "Google Image" with the key words "X-ray image". Figure demonstrates the effectiveness of four chosen method for the restoration of images blurred by uniform linear motion. These methods include: methods based on calculating the Moore–Penrose inverse, Wiener filter's method, Constrained least-squares (LS) filter's method, and the Lucy-Richardson methods, respectively. For the implementation of the last three method, we used built-in functions in Matlab software package. Legend with the tag original image is the original image divided at r = 750 rows and m = 1050 columns. To prevent the loss of valuable information from the edge of the image we have added zero boundary conditions, which implies zero value (black) to add pixels. This choice is a natural for the X-ray images simply because the background of the image is always black. Degraded image represents the image degradation obtained for l = 90. It is clear from Figure 3.2 depicting four methods to reconstruct blurred image in a way that details visible and clear.

The difference in the quality reconstructed image is difficult and not noticeable to the naked eye. Therefore, use common methods for comparing the restored files, or the so-called ISNR (improved signal-to-noise ratio).

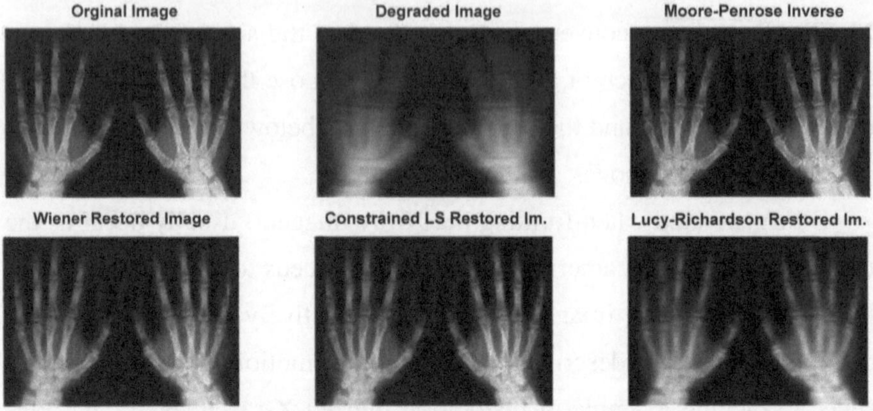

CPU time required to restore images to a method based on Moore–Penrose inverse, are shown in Figure 3.3 (right) and is provided in the function the parameter l. It is more than clear domination of the method is based on the block-partitioning. We will show that this method is not limited to use in X-ray. Funny, but that can be very successfully used in other practical problems. One of them is the problem of eliminating blur in images that appear in the automatic recognition of license panel (Automatic Number Plate Recognition-ANPR) vehicles. It is natural to assume that the blurring that occurs in the ANPR system caused by vertical movement by movement of the vehicle.

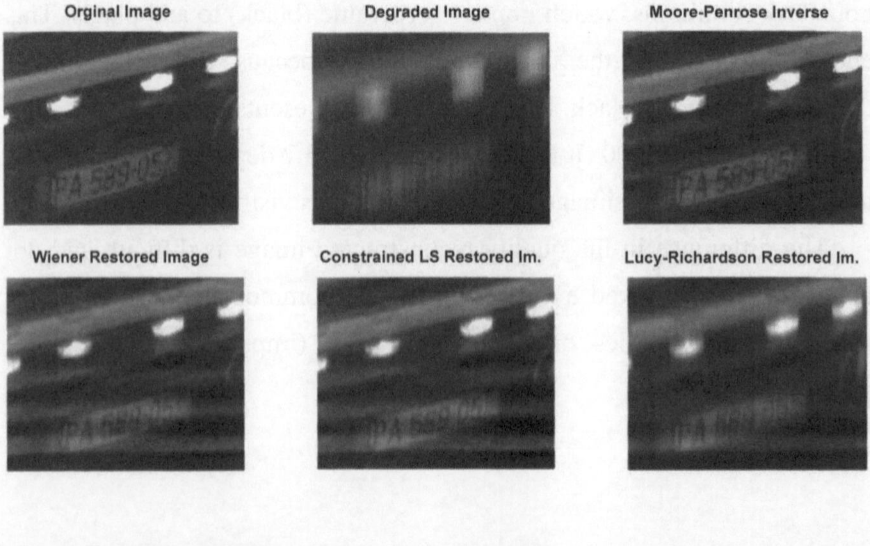

Figure 3.3 is the result of the restoration of the image dimensions of 1023×1250 taken from the ANPR system. The original picture was taken from the Customs Administration and the Republic of Macedonia was obtained by system for automated recognition of vehicle registration panel. In this case even the bare eye can note dominant method (restored image clarity) based on inverse MP in comparison to other methods.

CHAPTER 5

CONCLUSION

In this book, two groups of methods are analyzed:

1. Symbolic calculations of decomposition and generalized inverses of rational and polynomial matrix;
2. Generalized inverses of constant matrices.

The main objective of this book is a detailed analysis of the algorithmic approach to codes symbolic calculations wide class of generalized inverse constant, rational and polynomial matrices. The fact is that these areas are still unexplored and that there is room for further research in these fields. Known literature is usually dedicated to the theory of generalized inverses. In this topic plan application, the connection between the two disciplines is pointed out. In fact, it is possible to observe the theory and methods of unconditional optimization, as a useful tool for determining the generalized inverse matrix, and in some cases, for finding generalized inverse linear limited operator on Hilbert spaces. Later, it is possible to use the results presented here for the construction of new algorithms for solving specific optimization problems, to determine the different types of generalized inverse.

The aim of this study was to contribute both in theoretical (scientific) as well as the practical improvement existing methods and algorithms for effective symbolic matrix computations. In order to achieve these objectives in this book, special attention was paid to the following issues:

1. finding new and improvement of existing methods for the decomposition of polynomial and rational matrices;

2. the design of efficient algorithms for calculating the generalized inverse matrix;

3. comparison of new methods and algorithms with existing and displaying enhanced and promoted in terms of the complexity of oneness, the speed and efficiency;

4. use of the above-mentioned method for calculating the different types of symbolic generalized inverse polynomial and rational matrices;

In this, the concluding head is a brief overview of the results set forth, as well as some ideas and motivation for further research.

Implemented an efficient algorithm for calculating the $\{1,2,3\}$, $\{1,2,4\}$ and the inverse of Moore–Penrose inverse of-date rational matrix based on LDL^* factorization. This algorithm is very suitable for implementation in symbolic programming package `MATHEMATICA`. However, starting the algorithm is not suitable for use in procedural programming languages, because difficulties with the simplification of rational expressions. For this reason they made two algorithms for the calculation of Moore–Penrose-the inverse of polynomial and rational matrices. In them was used LDL^* full-rank factorization appropriate polynomial, rational or dies. It is avoided by the use of square roots that occur in the Cholesky decomposition.

It is known that the Gaussian elimination algorithm is not optimal. Even if it is invertible polynomial matrices, standard procedure of elimination may have problems with efficiency. Long-lasting normalization procedure must be performed (for example, determination of whether a pivot element equal to zero, or at cancelation). The degree of the numerator and denominators among-the results can grow exponentially from n, and hence the algorithm has exponential complexity. It can be reduced to a polynomial time complexity in applying the Gauss-Bareiss's technique, and even to a lesser degree by using interpolation techniques and reconstruction of rational functions.

Gauss algorithm with fractions is introduced in Ref. [2], which is a large improvement over the Gaussian elimination algorithm, and that performs all the steps where the exact share (i.e., without any residues) possible. Several recent improvement reduce the complexity of most problems with polynomial matrices in the order of magnitude of complexity of matrix multiplication.

One of the motivations for future research is the creation of similar algorithms, based on LDL^* or LDU decomposition, to work with the poor conditioned matrices. Also, it is necessary to find modifications introduced algorithms on some structural Hermitian matrices.

Numerical algorithm for calculation of $A^{(2)}_{T,S}$ inversion was carried out in Ref. [60], which generalizes the famous representation of the Moore–Penrose's inverse $A^\dagger$ from Ref. [33]. The implemented algorithm is based on calculating the QR decomposition full-rank matrix corresponding matrix W.

It is proved that the obtained external representation inverse and corresponding general representation based on an arbitrary factorization full-rank produce identical results. Was investigated and explicit transitional formula between these representations.

Also, special complete factorization rank that generate $\{2,4\}$ and $\{2,3\}$-inverses are obtained. All algorithms have been introduced based on the QR decomposition are highly efficient compared to other methods for calculating equalization Moore–Penrose's and Drazin inverse.

Two algorithms are obtained for symbolic calculation of $A_{T,S}^{(2)}$ rational inverse matrix of a variable. We watched the symbolic calculation using the generalized inverse LDL^* and QDR decomposition complete ranking fixed matrix W. By using additional matrices D in both cases, avoided the use of the square root, which is of crucial symbolic codes calculating polynomial and rational expressions. Data are comparative computation times of the algorithms are provided, along with Leverrier–Faddeev algorithm and partitioning method.

Described is a method to block LDL^* decomposition full-rank constant matrix block. This method has been applied to finding the inverse and Moore–Penrose's inverse of 2×2 block matrix. Based on initial results, a method was developed for the direct determination directly sub-matrices that appear in the Moore–Penrose's inverse. It has been shown that these methods are suitable for the implementation in procedural languages.

Expanding and generalizing these results to the case of rational matrices and for the case of two variables represents motivation for further research. Also, other types of generalized inverse can be determined using the block LDL^* decomposition of the full-rank.

BIBLIOGRAPHY

[1] Akritas, A. G., Malaschonok, G. I. (2007). Computations in modules over commutative domains, *Lecture Notes in Computer Science*, **4770**, 11–23.

[2] Bareiss, E. H. (1968). Sylvester's identity and multistep integer-preserving Gaussian elimination, *Mathematics of Computation*, **22**, 565–578.

[3] Ben-Israel, A., Greville, T. N. E. (2003). *Generalized Inverses, Theory and Applications, Second edition*, Canadian Mathematical Society, Springer, New York.

[4] Bhatti, M. A. (2000). *Practical Optimization Methods with Mathematica® Applications*, Springer-Verlag, New York

[5] Campbell, S. L., Meyer, Jr. C. D. (2009). *Generalized inverse of linear transformation*, Pitman, London (1979), SIAM.

[6] Chen, Y. (1990). The generalized Bott–Duffin inverse and its application, *Linear Algebra Appl.*, **134**, 71–91.

[7] Chen, Y., Chen, X. (2000). Representation and approximation of the outer inverse $A_{T,S}^{(2)}$ of a matrix A, *Linear Algebra Appl.* **308**, 85–107.

[8] Chow, E., Saad, Y. (1997). Approximate inverse techniques for block-partitioned matrices, *SIAM J. Sci. Comput.* **18**, 1657–1675.

[9] Cline, R. E. (1968). Inverses of rang invariant powers of a matrix, *SIAM J. Numer. Anal.* **5**, 182–197.

[10] Cormen, T. H., Leiserson, C. E., Rivest, R. L., Stein, C. (2001). *Introduction to Algorithms, Second Edition*, The MIT Press, Cambridge, Massachusetts London, McGraw-Hill Book Company, Boston, New York, San Francisco, St. Louis, Montreal, Toronto.

[11] Courrieu, P. (2005). Fast computation of Moore-Penrose inverse matrices, *Neural Information Processing – Letters and Reviews* **8**, 25–29.

[12] Courrieu, P. (2009). Fast solving of weighted pairing least-squares systems, *Journal of Computational and Applied Mathematics* **231**, 39–48.

[13] Cvetković-Ilić, D. S. (2008). A note on the representation for the Drazin inverse of 2×2 block matrices, *Linear Algebra and its Applications* **429**, 242–248.

[14] Cvetković-Ilić, D. S. (2012). New conditions for the reverse order laws for $\{1,3\}$ and $\{1,4\}$-generalized inverses, *Electronic Journal of Linear Algebra*, **23**, 231–242.

[15] Cvetković-Ilić, D. S. (2011). New additive results on Drazin inverse and its applications, *Appl. Math. Comp.* **218** (7), 3019–3024.

[16] Cvetković-Ilić, D. S., Stanimirović, P., Miladinović, (2011). Comments on some recent results concerning $\{2,3\}$ and $\{2,4\}$-generalized inverses, *Appl. Math. Comp.* **217** (22), 9358–9367.

[17] Fragulis, G., Mertzios, B. G., Vardulakis, A. I. G., (1991). Computation of the inverse of a polynomial matrix and evaluation of its Laurent expansion, *Int. J. Control*, **53**, 431–443.

[18] Giorgi, P., Jeannerod, C. P., Villard, G. (2003). On the Complexity of Polynomial Matrix Computations, *Proceedings of the 2003 International Symposium on Symbolic and Algebraic Computation*, 135–142.

[19] Golub, G. H., Van Loan, C. F. (1996). *Matrix Computations, Third edition*, The Johns Hopkins University Press, Baltimore.

[20] Goodall, C. R. (1993). Computation Using the QR Decomposition, C. R. Rao, ed., *Handbook of Statistics*, bf 9, 467–508.

[21] Greville, T. N. E. (1960). Some applications of the pseudo-inverse of matrix, *SIAM Rev.* **3**, 15–22.

[22] Getson, A. J., Hsuan, F. C. (1988). $\{2\}$-Inverses and their Statistical Applications, *Lecture Notes in Statistics* **47**, Springer, Berlin.

[23] Hartwig, R. E. (1981). A method for calculating A^d, *Math. Japonica* **26**, 37–43.

[24] Higham, N. J. *The Matrix Computation Toolbox*, http://www.ma.man.ac.uk/~higham/mctoolbox.

[25] Higham, N. J. (2002). *Accuracy and Stability of Numerical Algorithms*, Second Edition, SIAM, Philadelphia.

[26] Hung, C., Markham, T. L. (1975). The Moore-Penrose Inverse of a Partitioned Matrix $M = \begin{pmatrix} AD \\ BC \end{pmatrix}$, *Linear Algebra and Its Applications*, **11**, 73–86.

[27] Husen, F., Langenberg, P., Getson, A. (1985). The $\{2\}$-inverse with applications to statistics, *Linear Algebra Appl.* **70**, 241–248.

[28] Jeannerod, C. P., Villard, G. (2005). Essentially optimal computation of the inverse of generic polynomial matrices, *Journal of Complexity* **21**, 72–86.

[29] Jones, J., Karampetakis, N. P., Pugh, A. C. (1998). The computation and application of the generalized inverse via Maple, *J. Symbolic Computation* **25**, 99–124.

[30] Karampetakis, N. P. (1997). Computation of the generalized inverse of a polynomial matrix and applications, *Linear Algebra Appl.* **252**, 35–60.

[31] Karampetakis, N. P., Tzekis, P. (1998). On the computation of the generalized inverse of a polynomial matrix, *6th Medit. Symposium on New Directions in Control and Automation*, 1–6.

[32] Karampetakis, N. P. (1997). Generalized inverses of two-variable polynomial matrices and applications, *Circuits Systems Signal Processing* **16**, 439–453.

[33] Katsikis, V. N., Pappas, D., Petralias, A. (2011). An improved method for the computation of the Moore-Penrose inverse matrix, *Appl. Math. Comput.* **217**, 9828–9834.

[34] Lu, T., Shiou, S. (2002). Inverses of 2×2 block matrices, *Computers and Mathematics with Applications* **43**, 119–129.

[35] Maeder, R. (1996). *Programming in Mathematica, Third Edition* Redwood City, California: Adisson-Wesley.

[36] Malyshev, A. N. (1995). "Matrix equations: Factorization of matrix polynomials" M. Hazewinkel (ed.), *Handbook of Algebra I*, Elsevier, 79–116.

[37] Miettien, K. (1999). *Nonlinear Multiobjective Optimization*, Kluver Academic Publishers, Boston, London, Dordrecht.

[38] Mulders, T., Storjohann, A. (2003). On lattice reduction for polynomial matrices, *J. Symbolic Comput.* **35** (4) 377–401.

[39] Matrix Market, *National Institute of Standards and Technology*, Gaithersburg, MD. Available online from: http://math.nist.gov/MatrixMarket.

[40] Nagy, L., Miller, V., Powers, D. (1976). Research note: on the application of matrix generalized inverses to thc construction of inverse systems, *Int. J. Control*, **24**, 733–739.

[41] Najafi, H. S., Solary, M. S. (2006). Computational algorithms for computing the inverse of a square matrix, quasi-inverse of a non-square matrix and block matrices, *Appl. Math. Comput.* **183**, 539–550.

[42] Nashed, M. Z. (1976). *Generalized Inverse and Applications*, Academic Press, New York.

[43] Nashed, M. Z., Chen, X. (1993). Convergence of Newton-like methods for singular operator equations using outer inverses, *Numer. Math.* **66**, 235-257.

[44] Noble, B. (1966). A method for computing the generalized inverse of a matrix, *SIAM J. Numer. Anal.* **3**, 582–584.

[45] Petković, M. D., Stanimirović, P. S., Tasić, M. B. (2008). Effective partitioning method for computing weighted Moore–Penrose inverse, *Comput. Math. Appl.* **55**, 1720–1734.

[46] Petković, M. D., Stanimirović, P. S. (2005). Symbolic computation of the Moore-Penrose inverse using partitioning method, *Int. J. Comput. Math.* **82**, 355–367.

[47] Piziak, R., Odell, P. L. (1999). *Full rang Factorization of Matrices*, Mathematics Magazine **72**, 193–201.

[48] Rao, C. R., Mitra, S. K. (1971). *Generalized Inverse of Matrices and its Applications*, John Wiley & Sons, Inc., New York, London, Sydney, Toronto.

[49] Rakha, M. A. (2004). On the Moore-Penrose generalized inverse matrix, *Appl. Math. Comput.* **158**, 185–200.

[50] Robert, P. (1968). On the Group inverse of a linear transformation, *J. Math. Anal. Appl.* **22**, 658–669.

[51] Rodman, L. (1995). "Matrix Functions" M. Hazewinkel (ed.), *Handbook of Algebra*, Elsevier, 117–154.

[52] Schittkowski, K. (2004). *Multicriteria Optimization: User's Guide*, http://www.klaus-schittkowski.de, November.

[53] Sheng, X., Chen, G. (2007). Full-rang representation of generalized inverse $A_{T,S}^{(2)}$ and its applications, *Comput. Math. Appl.* **54**, 1422–1430.

[54] Shinozaki, N., Sibuya, M., Tanabe, K. (1972). Numerical algorithms for the Moore-Penrose inverse of a matrix: direct methods, *Annals of the Institute of Statistical Mathematics* **24**, 193–203.

[55] Stadler, W. (1988). Fundamentals of multicriteria optimization, In: Stadler, W.(ed.) *Multicriteria Optimization in Engineering and in the Sciences*, pp.1–25.New York: Plenum Press.

[56] Stanimirović, I. P. Computing Moore-Penrose inverse using full-rank square root free factorizations of rational matrices, submitted to *Journal of Computational Mathematics*.

[57] Stanimirović, I. P. (2012). Full–rank block LDL* decomposition and the inverses of $n \times n$ block matrices, *Journal of Applied Mathematics and Computing* **40**, 569–586.

[58] Stanimirović, I. P., Tasić, M. B. (2012). Computation of generalized inverses by using the LDL* decomposition, *Appl. Math. Lett.*, **25**, 526–531.

[59] Stanimirović, I. P., Tasić, A. M. Ilić, M. B. Full-rank LDL* Decomposition and Generalized Inverses of Polynomial Hermitian Matrices, submitted to *Journal of Symbolic Computations*.

[60] Stanimirović, P., Pappas, D., Katsikis, V., Stanimirović, I. (2012). Full–rank representations of outer inverses based on the QR decomposition, *Applied Mathematics and Computation* **218**, 10321–10333.

[61] Stanimirović, P., Pappas, D., Katsikis, V., Stanimirović, I. (2012). Symbolic computation of $A^{(2)}_{T;S}$-inverses using QDR factorization, *Linear Algebra Appl.* **437**, 1317–1331.

[62] Stanimirović, P. S. (1998). Block representations of $\{2\}, \{1,2\}$ inverses and the Drazin inverse, *Indian J. Pure Appl. Math.* **29**, 1159–1176.

[63] Stanimirović, P. S. Tasić, M. B. (2001). Drazin Inverse of One-Variable Polynomial Matrices, *Filomat*, **15**, 71–78.

[64] Stanimirović, P. S., Tasić, M. B., Krtolica, P., Karampetakis, N. P. (2007). Generalized inversion by interpolation, *Filomat*, **21** (1), 67–86.

[65] Stanimirović, P. S., Tasić, M. B. (2004). Partitioning method for rational and polynomial matrices, *Appl. Math. Comput.* **155**, 137–163.

[66] Stanimirović, P. S., Tasić, M. B. (2011). On the Leverrier-Faddeev algorithm for computing the Moore-Penrose inverse, *J. Appl. Math. Comput.* **35**, 135–141.

[67] Stanimirović, P. S., Djordjević, D. S. (2000). Full-rank and determinantal representation of the Drazin inverse, *Linear Algebra Appl.* **311**, 31–51.

[68] Stanimirović, P. S., Cvetković-Ilić, D. S., Miljković, S., Miladinović, M. (2011). Full-rank representations of $\{2,4\}, \{2,3\}$-inverses and successive matrix squaring algorithm, *Appl. Math. Comput.* **217**, 9358–9367.

[69] Stanimirović, P. S., Tasić, M. B. (2008). Computing generalized inverses using LU factorization of matrix product, *Int. J. Comput. Math.* **85**, 1865–1878.

[70] Stojohann, A., Labahn, G. (1997). A Fast Las Vegas algorithm for Computing the Smith Normal Form of a Polynomial Matrix, *Linear Algebra Appl.* **253**, 155–173.

[71] Tasić, M., Stanimirović, I., (2010). Implementation of partitioning method, Facta *Universitatis (Niš) Ser. Math. Inform.* **25**, 25–33.

[72] Tasić, M. B., Stanimirović, I. P., Symbolic computation of Moore-Penrose inverse using the LDL* decomposition of the polynomial matrix, *Filomat*, accepted for publication.

[73] Tasić, M., Stanimirović, P., Stanimirović, I., Petković, M., Stojković, N. (2005). Some useful MATHEMATICA teaching examples, *Facta Universitatis (Niš) Ser.: Elec. Energ.* **18**, No.2, 329–344.

[74] Tasić, M. B., Stanimirović, P. S., Petković, M. D. (2007). Symbolic computation of weighted Moore-Penrose inverse using partitioning method, *Appl. Math. Comput.* **189**, 615–640.

[75] Tasić, M. B., Stanimirović, P. S. (2008). Symbolic and recursive computation of different types of generalized inverses, *Appl. Math. Comput.* **199**, 349–367.

[76] Tasić, M. B., Stanimirović, P. S. (2010). Differentiation of generalized inverses for rational and polynomial matrices, *Appl. Math. Comput.* **216**, 2092–2106.

[77] Tewarson, R. P. (1967). A direct method for generalized matrix inversion, *SIAM J. Numer. Anal.* **4**, 499–507.

[78] Tian, Y. (1998). The Moore-Penrose inverses of $m \times n$ block matrices and their applications, *Linear Algebra and its Applications*, **283**, 35–60.

[79] Toutounian, F., Ataei, A. (2009). A new method for computing Moore-Penrose inverse matrices, *Journal of Computational and Applied Mathematics*, **228**, 412–417.

[80] Wang, G., Wei, Y., Qiao, S. (2004). *Generalized Inverses: Theory and Computations*, Science Press, Beijing/New York.

[81] Wang, G. R., Chen, Y. L. (1986). A recursive algorithm for computing the weighted Moore-Penrose inverse $A^{\dagger}_{MN}$, *J. Comput. Math.* **4**, 74–85.

[82] Watkins, D. (2002). *Fundamentals of Matrix Computations*, Wiley-Interscience, New York.

[83] Wei, Y. (1998). A characterization and representation of the generalized inverse $A^{(2)}_{T,S}$ and its applications, *Linear Algebra Appl.* **280**, 87–96.

[84] Wei, Y., Wu, H. (2003). The representation and approximation for the generalized inverse $A^{(2)}_{T,S}$, *Appl. Math. Comput.* **135**, 263–276.

[85] Winkler, F. (1996). *Polynomial Algorithms in Computer Algebra*, Springer.

[86] Wolfram, S. (2003). *The MATHEMATICA Book, 5th ed.*, Wolfram Media/Cambridge University Press, Champaign, IL 61820, USA.

[87] Zheng, Bapat, B. R. B. (2004). Generalized inverse $A^{(2)}_{T,S}$ and a rank equation, *Appl. Math. Comput.* **155(2)**, 407–415.

[88] Zielke, G. (1986). Report on test matrices for generalized inverses, *Computing* **36**, 105–162.

[89] Zielke, G. (1984). A survey of generalized matrix inverses, *Computational Mathematics*, Banach Center Publications **13**, 499–526.

[90] Zotos, K. (2006). Performance comparison of Maple and Mathematica, *Appl. Math. Comput.*, doi:10.1016/j.amc.2006.11.008.

[91] Zlobec, S. (1970). An explicit form of the Moore-Penrose inverse of an arbitrary complex matrix, *SIAM Rev.* **12**, 132–134.

INDEX

Q

R